Fundamentals of Plasticity in Geomechanics

Fundamentals of Plasticity in Geomechanics

S. Pietruszczak
McMaster University, Hamilton, Ontario, Canada

CRC Press
Taylor & Francis Group
Boca Raton London New York Leiden

CRC Press is an imprint of the
Taylor & Francis Group, an **informa** business
A BALKEMA BOOK

Published by: CRC Press/Balkema
P.O. Box 447, 2300 AK Leiden, The Netherlands
e-mail: Pub.NL@taylorandfrancis.com
www.crcpress.com – www.taylorandfrancis.co.uk – www.balkema.nl

First issued in paperback 2020

ISBN-13: 978-0-367-57714-8 (pbk)
ISBN-13: 978-0-415-58516-3 (hbk)

Visit the Taylor & Francis Web site at
http://www.taylorandfrancis.com

and the CRC Press Web site at
http://www.crcpress.com

British Library Cataloguing in Publication Data
A catalogue record for this book is available from the British Library

Library of Congress Cataloging-in-Publication Data

Pietruszczak, S.
Fundamentals of plasticity in geomechanics / S. Pietruszczak.
p. cm.
Includes bibliographical references.
ISBN 978-0-415-58516-3 (harcover : alk. paper)
1. Soils–Plastic properties. 2. Continuum mechanics. I. Title.

TA710.5.P54 2010
624.1'5136–dc22 2010013638

Typeset by Macmillan Publishing Solutions, Chennai, India

Contents

Preface

This book is based on a series of graduate lectures that I have been giving at McMaster University in Hamilton, Ontario, Canada. The book presents a simple, concise and reasonably comprehensive introduction to fundamental concepts of plasticity in relation to geomechanics. The text is intended primarily for Ph.D./M.Sc. students as well as researchers working in the areas of soil/rock mechanics. It may also be of interest to practicing engineers familiar with established notions of contemporary continuum mechanics.

The content of this book is divided into eight chapters. Chapter 1 gives a brief overview of the basic concepts and fundamental postulates. Chapter 2 presents a review of the elastic-perfectly plastic formulations in geomechanics. It examines the typical failure criteria for geomaterials and presents a general procedure for specification of the constitutive operator. Chapter 3 focuses on isotropic strain-hardening formulations. The presentation in this chapter is different, in terms of scope, from that in other available texts. The discussion includes the notions of volumetric, deviatoric, as well as combined volumetric-deviatoric hardening. Chapter 4 covers the isotropic-kinematic hardening rules formulated within the context of bounding surface plasticity. The framework outlined in this chapter provides an extension of classical isotropic hardening descriptions to deal with the case of cyclic/fluctuating loads; in particular, with the notion of irreversibility of deformation on stress reversals. This chapter is more advanced and may be omitted at the first reading. Chapter 5 outlines the basic techniques for numerical integration, while Chapter 6 gives an overview of procedures for limit analysis that include applications of lower and upper bound theorems. Both these chapters are introductory in nature and are intended to provide a basic background in the respective areas. Chapter 7 deals with description of inherent anisotropy in geomaterials. Two different phenomenological frameworks are outlined; one known as the critical plane approach and the other incorporating mixed invariants of stress and microstructure tensors. This chapter may again be omitted at the first reading as it is intended for a more advanced reader. Finally, Chapter 8 provides an overview of the experimental response of geomaterials. It supplements the theoretical considerations in earlier Chapters 3, 4 and 7 by examining the basic trends associated with mechanical response of both soil and rock-like materials.

The primary objective of this work is to provide the reader with a general background in soil/rock plasticity. As mentioned earlier, I tried to be both concise and yet comprehensive in the presentation of the material. At the same time, however, several important topics have not been covered. Those include the description of localized deformation, modeling of partially saturated soil, etc. Thus, the book should be perceived as an introduction to the broad area of inelastic response of geomaterials.

A difficult aspect in writing a textbook is to properly acknowledge the contribution of all researchers in the field. This is particularly complex in the context of soil/rock plasticity, where so many prominent scientists have made significant contributions. I have, in fact, purposely tried to limit the number of references so that not to distract the reader. I would, therefore, like to extend my apologies to all those researchers whose work I have not referenced; it did not stem from disregard, but rather from a need to be concise.

Finally, I would like to express my gratitude to Prof. D. Lydzba (Technical University of Wroclaw, Poland) and Prof. J.F. Shao (Lille University of Science and Technology, France) who were instrumental in getting this project started and contributed considerably to Chapters 5 and 6. I am also grateful to my close friend, Prof. G.N. Pande (Swansea University, U.K.), for reading the manuscript and providing valuable comments. Last but not the least, I wish to thank my family, particularly my wife Maria and my daughter Catherine, for their patience and understanding; I do appreciate it.

S. Pietruszczak
Hamilton, Ont.
March 2010

Chapter 1

Basic concepts of the theory of plasticity

The theory of plasticity was originally developed to describe the mechanical behaviour of metals beyond the elastic regime. The first attempt at a concise mathematical formulation was made as early as 1870 by S. de Venant. It was closely linked with theoretical studies of M. Levy and experimental investigations by H. Tresca. A comprehensive historical background on the main advances can be found, for example, in Refs. [1–3].

The first significant attempt at adopting the notions of metal plasticity to geomaterials was made in the early 1950's by Drucker and Prager, who extended the classical form of Coulomb criterion to address a general 3D case. The subsequent two decades brought a rapid development in the context of formulation of a broad range of strain-hardening concepts, including the description of plastic deformation on stress reversals. Again, a historical background on major advances can be found, for example, in Refs. [4,5].

This chapter provides a brief review of the fundamentals of the theory of plasticity. It is perceived as a concise introduction to subsequent chapters which address specific approaches for modeling of the mechanical response of geomaterials. Here, the basic conceptual framework is outlined first, followed by a review of fundamental postulates, such as stability of the material, uniqueness of the solution, etc.

1.1 TYPICAL APPROXIMATIONS OF UNIAXIAL RESPONSE OF MATERIALS

Consider a material whose quasi-static response under uniaxial stress conditions, i.e. in simple compression/tension, is as shown in Figure 1.1a. The trends depicted in this figure may be considered as typical for a broad class of engineering materials. For stress magnitudes σ satisfying $\sigma < \sigma_Y$ the response is assumed to be elastic, i.e. the strains are fully recoverable when the load is removed. For stress magnitudes exceeding σ_Y the material becomes elastic-plastic. The basic distinction between elasticity and plasticity is the irreversibility of plastic action. When the load is removed at A, the material accumulates certain irreversible strain ε^p and it is unable to resume its original shape. This means that the work done is not recoverable and that a plastically deformed body may be restored to its original shape only by an additional plastic action. For $\sigma_Y < \sigma < \sigma_U$ the plastic deformation is said to be stable, meaning that the stress is a monotonically increasing function of strain. At the instant the ultimate stress σ_U is reached, a transition to unstable behaviour may take place.

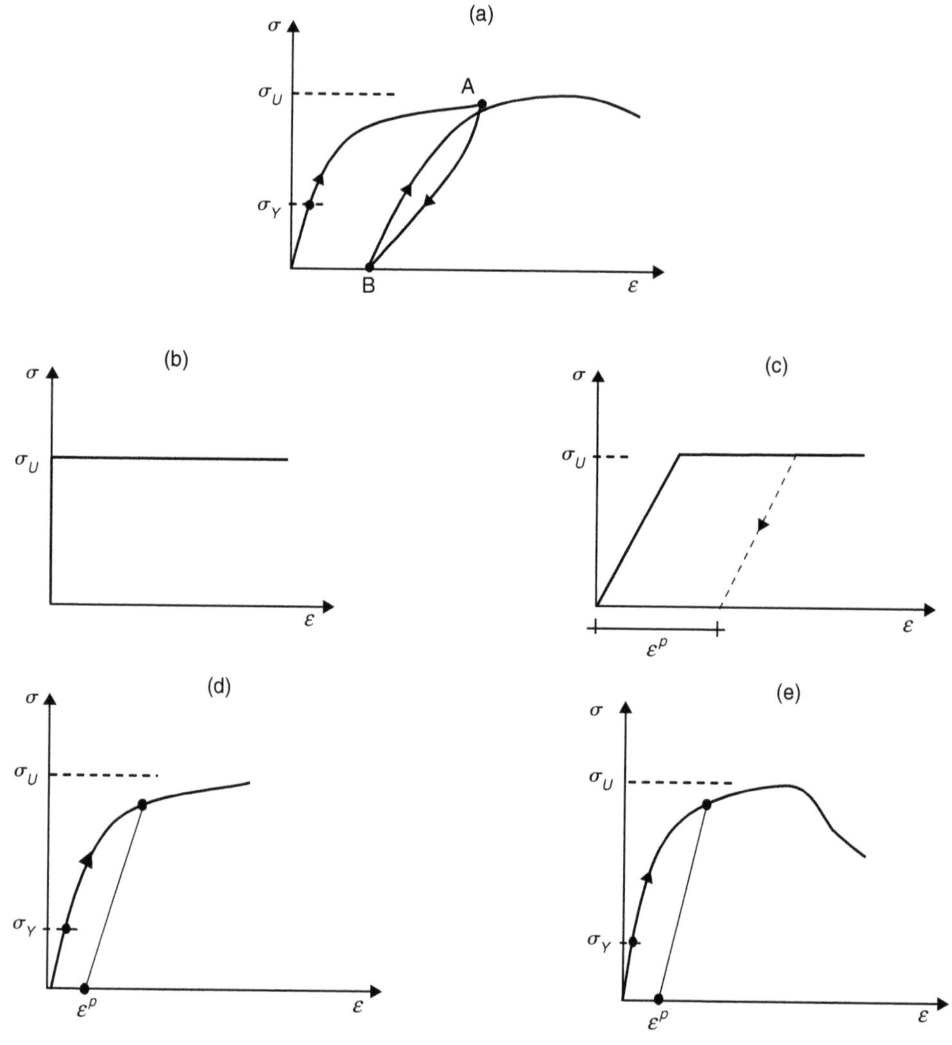

Figure 1.1 (a) A typical stress-strain characteristic in uniaxial case; (b) rigid-perfectly plastic; (c) elastic – perfectly plastic; (d) strain hardening; (e) strain hardening/softening idealizations

The purely elastic response is restricted to $\sigma < \sigma_Y$. For a limited class of materials (e.g. metals) σ_Y, referred to as yield stress, can be identified in a unique manner. However, in certain geomaterials, especially those naturally occurring such as soil, $\sigma_Y \to 0$ which implies that the irreversible deformation develops virtually from the beginning of the loading process.

In the theory of plasticity the stress-strain curves are usually approximated. Typical approximations are shown in Figure 1.1. The simplest of all is the concept of a *rigid-perfectly plastic* material, Figure 1.1b. Reversible deformations are completely neglected. When the ultimate stress intensity σ_U is reached the material becomes perfectly-plastic, i.e. unlimited plastic deformation takes place under $\sigma = \sigma_U$. In mathematical terms, the stress-strain relation corresponding to this approximation is not unique. For a prescribed stress $\sigma = \sigma_U$ the strain in the plastic range is indeterminate and conversely, for a prescribed strain $\varepsilon = 0$ the stress in $\sigma < \sigma_U$ range is not unique. The rigid-plastic idealization is used mainly in the stability (limit) analysis whose objective is to estimate the collapse load of a structure without an explicit reference to the actual deformation field (Chapter 6). The deformation history can be accounted for by considering the material as an *elastic-perfectly plastic*, Figure 1.1c. Although, in the elastoplastic range, the strain is still indeterminate from stress, in general however there is a locally unique response in stress for a given strain. The elastic behaviour is usually considered to be linear.

The concept of perfect plasticity has quite apparent limitations and a better approximation may be obtained by accounting for *strain-hardening*. In the one-dimensional context, strain-hardening means that the initial yield stress σ_Y is raised in the direction of deformation as a result of progressive accumulation of plastic strain. Referring to Figure 1.1d, the initial response is assumed to be elastic for all $\sigma < \sigma_Y$, whereas for $\sigma_Y < \sigma < \sigma_U$ the irreversible deformation takes place. If the material is loaded to, say, σ'_Y and subsequently unloaded, the yield stress is raised from σ_Y to σ'_Y, so that the elastic domain becomes $\sigma < \sigma'_Y$. The strain-hardening response always results in a unique and stable stress-strain characteristic.

Finally, some of the constitutive concepts for geomaterials incorporate also the effect of strain-softening, Figure 1.1e. The term softening implies an unstable response, i.e. stress is said to be a monotonically decreasing function of strain. It should be noted, that although the stress-strain mapping is not unique, Figure 1.1e, there is always a unique response in the incremental sense, i.e. for any consistently prescribed stress increment the corresponding strain increment can be found (and vice-versa). The strain-softening phenomenon is often associated with a deformation mode which is inhomogeneous on a macroscale (e.g. formation of a shear band). Therefore, it can not be interpreted strictly as a material response.

The above discussed approximations are very restrictive as they apply only to a uniaxial state of stress. In the subsequent section, the fundamental concepts introduced above are generalized for complex loading histories involving all stress/strain components.

1.2 THE NOTION OF GENERALIZED YIELD/FAILURE CRITERION

In a one-dimensional context, the transition from elastic to elastoplastic response is said to take place whenever the stress state σ reaches a critical intensity σ_Y. Under an arbitrary stress state, the notion of the yield stress σ_Y is generalized by postulating the existence of a function f_o of the stress state σ_{ij}, i.e. $f_o(\sigma_{ij})$, such that if $f_o < 0$ the response is elastic, whereas for $f_o = 0$ the plastic deformation takes place in a body which is under a homogeneous stress. Thus, $f_o(\sigma_{ij}) = 0$ represents a generalized *yield criterion*, i.e. it defines the limit of elasticity under any possible combination of stress.

A similar methodology is employed to generalize the notion of the ultimate stress σ_U, Figure 1.1b, at which an unlimited plastic flow takes place. In this case, the existence of a function $F(\sigma_{ij})$ is assumed such that, if the ultimate state of the material is reached, the value of this function is constant, say zero. Thus, $F(\sigma_{ij})=0$ represents a generalized *failure criterion*, i.e. it defines the state at which unrestricted plastic deformation can take place in a body which is under a homogeneous stress.

Both $f_o(\sigma_{ij})=0$ and $F(\sigma_{ij})=0$ can be regarded as surfaces in a nine-dimensional Euclidean space in which stresses σ_{ij} are taken as nine rectangular Cartesian coordinates. Since σ_{ij} is symmetric, a similar interpretation in a six-dimensional subspace is also feasible. Let us now focus on the general representation of both these functions. For an isotropic material, i.e. one which has no directional properties in itself, f_o as well as F should assume the same values for all orientations of the Cartesian coordinate axes relative to the material, that is

$$f_o(\sigma_{ij}) = f_o(T_{ip}T_{jq}\sigma_{pq}); \quad F(\sigma_{ij}) = F(T_{ip}T_{jq}\sigma_{pq}) \tag{1.1}$$

where T_{ij} is the transformation tensor. The function satisfying (1.1) is referred to as isotropic with respect to σ_{ij} and according to the representation theorems (see, e.g., Refs. [6,7]) can be expressed in terms of the basic invariants of σ_{ij}

$$f_o(\sigma_{ij}) = f_o(I_1, I_2, I_3); \quad F(\sigma_{ij}) = F(I_1, I_2, I_3) \tag{1.2}$$

where $I_1 = \sigma_{ii}$, $I_2 = (\sigma_{ij}\sigma_{ij} - \sigma_{ii}\sigma_{jj})/2$, $I_3 = \det(\sigma_{ij})$. Alternatively, f_o and F may be a function of any convenient set of invariant measures which are derived from the basic invariants. In particular, since the stress invariants are completely defined by the magnitudes of the principal stresses $\sigma_1, \sigma_2, \sigma_3$, the representation (1.2) can be replaced by

$$f_o(\sigma_{ij}) = f_o(\sigma_1, \sigma_2, \sigma_3); \quad F(\sigma_{ij}) = F(\sigma_1, \sigma_2, \sigma_3) \tag{1.3}$$

Although any function of the type (1.2) can be expressed as (1.3) it is not true that an arbitrary function of the principal stresses is an admissible failure criterion. For an isotropic material f_o and F must be symmetric functions of $\sigma_1, \sigma_2, \sigma_3$, i.e. they must be such that if, say, $f_o(\sigma_1, \sigma_2, \sigma_3)=0$ then so does $f_o(\sigma_2, \sigma_1, \sigma_3)=0$, since the material is unable to distinguish how the principal stresses are labelled. Finally, it is noted that the functional form (1.1) implies that the conditions at the onset of yielding as well as those at failure are *path-independent*.

1.3 GENERALIZATION OF THE CONCEPTS OF PERFECTLY PLASTIC AND STRAIN-HARDENING MATERIAL

Consider first an elastic-perfectly plastic material, as depicted in Figure 1.2. The significance of perfect plasticity in uniaxial case is that the stress remains constant, $\sigma = \sigma_U$, as the strain increases. In the general context, it is postulated that during the plastic flow the state of stress always satisfies $F=0$. In other words, the only admissible stress trajectories are those for which $F=0$, whereas those resulting in $F>0$ are not

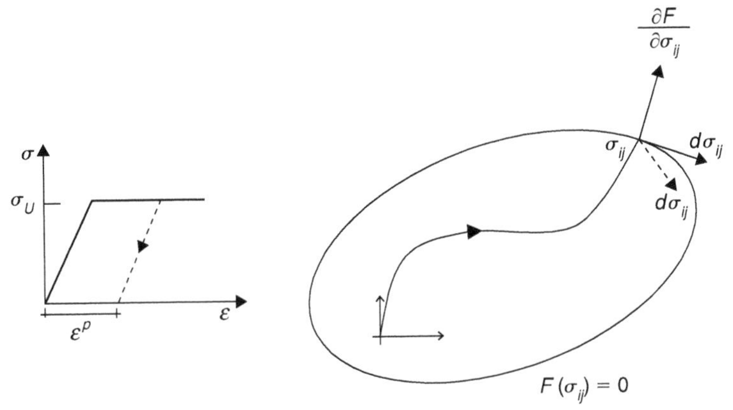

Figure 1.2 Elastic – perfectly plastic material; schematic representation

permitted. Consequently, since the surface $F=0$ is stationary, the differential stress increments, $d\sigma_{ij}$, must satisfy

$$F(\sigma_{ij})=0 \quad \Rightarrow \quad dF=\frac{\partial F}{\partial \sigma_{ij}}d\sigma_{ij}=0 \tag{1.4}$$

The above expression is referred to as the *consistency condition*. The geometric interpretation of eq.(1.4) is quite apparent. Since the gradient of F represents a vector along the outward normal to $F=0$, Figure 1.2, the stress increment must remain perpendicular to it, and thus tangential to the failure surface.

Equation (1.4) provides a general criterion for an *active loading* process in the plastic range. For all trajectories penetrating the domain enclosed by the failure surface, i.e.

$$F(\sigma_{ij})<0 \quad or \quad F=0 \quad \wedge \quad dF=\frac{\partial F}{\partial \sigma_{ij}}d\sigma_{ij}<0 \tag{1.5}$$

the response is said to be elastic. Note the second condition in (1.5) corresponds to an instantaneous unloading from $F=0$.

Consider now a strain-hardening idealization, Figure 1.3. The concept of the initial yield stress σ_Y can be generalized by introducing a yield criterion $f_o(\sigma_{ij})=0$ which represents the limit to elasticity under any possible combination of stress. Again, $f_o(\sigma_{ij})=0$ may be regarded as a surface in a nine/six-dimensional stress space.

In the uniaxial case, the yield stress σ_Y is progressively raised in the direction of deformation, i.e. stress is a monotonically increasing function of strain. For a combined state of stress/strain however, no such simple picture can be drawn. In this case, the plasticity generalization is based on assuming the existence of a *loading surface* $f=0$ which depends not only on the stress σ_{ij} but also on the plastic strain ε_{ij}^p,

$$f=f(\sigma_{ij},\varepsilon_{ij}^p)=0 \tag{1.6}$$

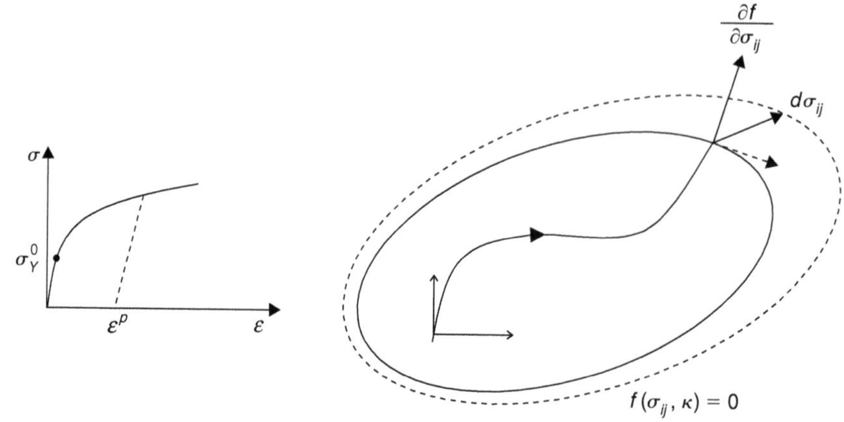

Figure 1.3 Isotropic strain – hardening; schematic representation

In eq.(1.6), the plastic strain seems to be more appropriate than the total strain, since the elastic contribution is uniquely determined by the stresses through the elasticity relation. The function f is such that for $\varepsilon^p_{ij}=0$ there is $f=f_o(\sigma_{ij})$, i.e. the loading surface coincides with the initial yield surface $f_o=0$. Obviously, in geomaterials with no apparent elastic limit $f_o \to 0$ for $\sigma_{ij} \to 0$, i.e. the initial yield surface reduces to a point. As the deformation progresses beyond the elastic limit, the loading surface undergoes evolution due to accumulated plastic strain.

The simplest, and the most frequently used, variant of generalization is the concept of *isotropic hardening*. The loading surface (1.6) is expressed as

$$f(\sigma_{ij},\kappa)=\bar{f}(\sigma_{ij})-g(\kappa)=0 \tag{1.7}$$

where κ is a scalar parameter, so-called *hardening parameter*, which depends on the history of plastic deformation, and $g(\kappa)$ is a monotonically increasing function of κ.

Figure 1.3 provides a schematic geometric interpretation of the concept of isotropic hardening. For stress trajectories penetrating the region inside the initial yield surface, i.e. those satisfying $f_o=\bar{f}(\sigma_{ij})-g(0)<0$, the response is elastic. When $f_o=0$ and the differential stress increment $d\sigma_{ij}$ is directed outside the domain enclosed by the loading surface, the plastic flow takes place. An increase in κ, and thus in $g(\kappa)$, results, according to eq.(1.7), in a uniform (isotropic) expansion of the loading surface. The state of stress $\sigma_{ij}+d\sigma_{ij}$ is located on the subsequent (updated) loading surface, so that the condition $f=0$ is satisfied throughout the deformation process. Thus, by analogy to the uniaxial case, the initial elastic limit is progressively 'raised' in the course of deformation.

The loading-unloading criteria can be formally derived from the consistency condition,

$$f(\sigma_{ij},\kappa)=0 \quad \Rightarrow \quad df=d\bar{f}-dg=\frac{\partial\bar{f}}{\partial\sigma_{ij}}d\sigma_{ij}-\frac{\partial g}{\partial\kappa}d\kappa=0 \tag{1.8}$$

which requires that the state of stress σ_{ij} be always located on the updated loading surface. Since during the plastic flow $dg > 0$, the *active loading* process is defined as

$$f = 0 \quad \wedge \quad d\bar{f} = \frac{\partial \bar{f}}{\partial \sigma_{ij}} d\sigma_{ij} > 0 \tag{1.9}$$

i.e., the stress increment is outside the domain contained within the current loading surface (so that $d\bar{f} = dg$). The case

$$f = 0 \quad \wedge \quad d\bar{f} = \frac{\partial \bar{f}}{\partial \sigma_{ij}} d\sigma_{ij} = 0 \tag{1.10}$$

represents a *neutral loading*. Since $df = 0$ implies $dg = 0$, no plastic deformation takes place and the loading surface remains stationary.

Finally, when $f = 0$ and $df < 0$, the consistency condition (1.8) can not be satisfied (in view of $dg > 0$) and the material undergoes unloading. Thus, the elastic response takes place whenever

$$f < 0 \quad or \quad f = 0 \quad \wedge \quad d\bar{f} = \frac{\partial \bar{f}}{\partial \sigma_{ij}} d\sigma_{ij} < 0 \tag{1.11}$$

The loading surface defined by eq.(1.7) always maintains its shape, while its isotropic expansion is governed by the hardening parameter κ. In geotechnical applications, κ is usually identified with the accumulated volumetric or distortional plastic strain

$$\kappa = \int d\varepsilon^p_{ii} \tag{1.12}$$

or

$$\kappa = \int d\varepsilon^p; \quad d\varepsilon^p = \left(\frac{1}{2} de^p_{ij}\, de^p_{ij}\right)^{1/2}; \quad de^p_{ij} = d\varepsilon^p_{ij} - \frac{1}{3}\delta_{ij} d\varepsilon^p_{kk} \tag{1.13}$$

or with the plastic work (so-called *work-hardening* approach)

$$\kappa = \int \sigma_{ij}\, d\varepsilon^p_{ij} \tag{1.14}$$

The integration in (1.12) through (1.14) is carried out over the entire plastic strain path.

An alternative way of describing the strain-hardening material is to incorporate the notion of *kinematic hardening*. In this case, the loading surface is assumed to undergo a continuing translation without the change in shape. Thus,

$$f = f(\sigma_{ij} - \alpha_{ij}) = 0; \quad \alpha_{ij} = \alpha_{ij}(\varepsilon^p_{kl}) \tag{1.15}$$

where α_{ij}, referred to as a *back stress*, specifies the position of the loading surface relative to the origin of the stress space. Now, the consistency condition reads

$$df = \frac{\partial f}{\partial \sigma_{ij}} d\sigma_{ij} + \frac{\partial f}{\partial \alpha_{ij}} d\alpha_{ij} = 0 \tag{1.16}$$

while the definitions of active and neutral loading are analogous to those employed in the case of isotropic hardening.

The simplest hypothesis for the translation of the loading surface is that due to Prager [8], which stipulates that $d\alpha_{ij} \propto d\varepsilon^p_{ij}$. An alternative formulation is due to Ziegler [9] and incorporates an assumption that direction of the translation is along $\sigma_{ij} - \alpha_{ij}$, i.e. $d\alpha_{ij} \propto (\sigma_{ij} - \alpha_{ij})$.

It is worth noting that the concept of kinematic hardening can account for certain manifestations of anisotropy induced by plastic flow. An example of this is provided in Figure 1.4, which presents a von Mises yield surface in the principal stress space $\sigma_1, \sigma_2, \sigma_3 = 0$ (plane stress). Initially, there is $\alpha_{ij} = 0$, so that $f = f_o(\sigma_{ij}) = 0$. In this case, uniaxial tension produces the same response in directions 1 and 2. Also, at this stage, the yield stress in tension is the same as that in compression. Consider now a uniaxial tension beyond the elastic limit, as shown in Figure 1.4a. The plastic deformation results in progressive translation of the yield surface and, as a result, the symmetry in behaviour no longer exists. In particular, the response along the directions 1 and 2 is different and the yield stress in compression is reduced as compared to that in tension (Figure 1.4b). This phenomenon, known as the *Bauschinger effect*, is common in metal plasticity [1].

It should be emphasized that the representation (1.15) is restrictive, as it does not incorporate any explicit measure of material fabric. Consequently, its ability to describe the anisotropy is quite limited. In fact, the primary feature of anisotropy, which involves the dependence of the response on the orientation of material triad in relation to principal stress axes, cannot be accounted for.

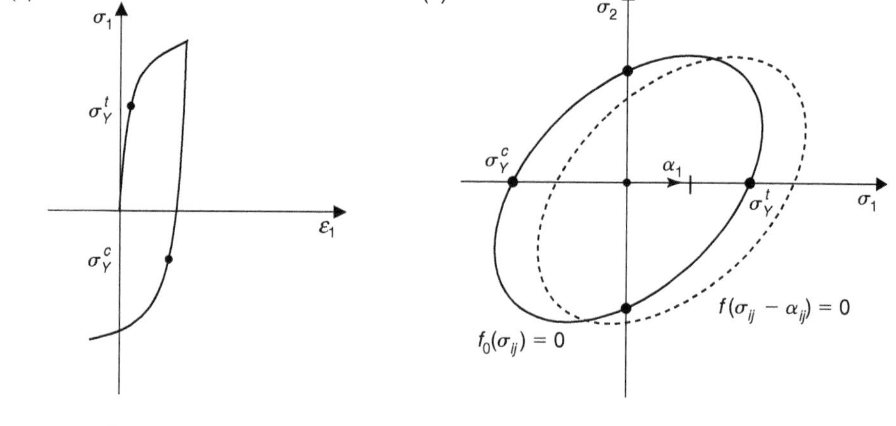

Figure 1.4 Kinematic hardening; representation of the Bauschinger effect

Finally, the description of mechanical response may invoke a combination of isotropic and kinematic hardening. In this case, the loading surface is defined as

$$f = \bar{f}(\sigma_{ij} - \alpha_{ij}) - g(\kappa) = 0 \tag{1.17}$$

i.e., it undergoes a simultaneous translation and uniform expansion.

1.4 DETERMINATION OF PLASTIC STRAIN; DEFORMATION AND FLOW THEORIES OF PLASTICITY

The classical theory of plasticity incorporates two types of relations for the evaluation of plastic strain. The first one, known as *deformation theory*, relates the total plastic strain to the history of stress. This approach, which was introduced in the early 1920's, is quite restrictive and it is primarily of a historical significance. A more common approach is the one which relates the plastic strain *increments* to the stress and stress increments. The latter is known as the *flow theory* of plasticity. In what follows, both these approaches are briefly reviewed and a general form of the constitutive relations is specified.

(i) Deformation theory

This simple approach was developed by Hencky [10] in the context of metal plasticity. The primary postulate involved here is that of coaxiality of plastic strain tensor and the stress deviator. Thus, in the plastic range

$$\varepsilon^p_{ij} = \beta(J_2)\, s_{ij} \tag{1.18}$$

where $\beta = \beta(J_2)$ is a scalar-valued function of J_2, i.e. the second invariant of the stress deviator s_{ij}. Clearly, $\varepsilon^p_{ii} = 0$, so that the plastic deformation is considered as incompressible, which is well documented for low porosity materials, like metals.

The behaviour in the elastic range is governed by Hooke's law, i.e.

$$e^e_{ij} = \frac{s_{ij}}{2G}; \quad \varepsilon^e_{ii} = \frac{\sigma_{kk}}{3K} = \varepsilon_{ii} \tag{1.19}$$

where G and K are the shear and bulk moduli, respectively. Assuming now additivity of elastic and plastic strain yields

$$e_{ij} = e^e_{ij} + e^p_{ij} = \left(\frac{1}{2G} + \beta\right) s_{ij} \tag{1.20}$$

which, in turn, implies that

$$e_{ij} = \Omega s_{ij}; \quad \Omega = \frac{1}{2G} + \beta \tag{1.21}$$

so that the strain and stress deviators remain coaxial.

The formulation, as outlined above, is clearly *path-independent*. If we move, in the stress space, from an arbitrary point O to, say, O_1 by two different trajectories, then the strain components at O_1 remain the same, which is enforced by the functional form (1.21). This is, in general, contradictory to the existing experimental evidence. Furthermore, for a neutral stress path there is no continuity between the elastic and plastic regions. The latter can be demonstrated by differentiating eq. (1.18)

$$d\varepsilon_{ij}^p = \beta(J_2)ds_{ij} + s_{ij}\, d\beta(J_2) \tag{1.22}$$

Apparently, if $J_2 = const.$ there is $d\beta = 0$, so that $d\varepsilon_{ij}^p = \beta(J_2)\, ds_{ij}$. Thus, the plastic strain does not vanish when the stress point moves along the envelope $J_2 = const.$

It is noted that the equations presented above are, in fact, the same as the equations governing the behaviour of a non-linear elastic body. The only difference here is a formal introduction of a loading criterion, as stipulated by the condition $dJ_2 > 0$.

(ii) Flow theory

The choice of possible relations is fairly restricted by the fact that, for strain-hardening material, the plastic strain increments should vanish for a *neutral loading path*. The latter, according to eq. (1.10), is defined as

$$d\bar{f} = \frac{\partial f}{\partial \sigma_{ij}}\, d\sigma_{ij} = 0 \tag{1.23}$$

Therefore,

$$d\varepsilon_{ij}^p = G_{ij}\, d\bar{f} \tag{1.24}$$

where G_{ij} is a symmetric second order tensor.

In the flow theory of plasticity, the specification of G_{ij} is based on an observation that in materials like metals the direction of plastic strain increment is not influenced by the stress increment. Consequently, for this class of materials

$$G_{ij} = G_{ij}(\sigma_{kl}, \kappa) \tag{1.25}$$

where κ is a scalar parameter which depends on the history of plastic deformation, viz. (1.12)–(1.14).

Given the representation (1.25) it is convenient now to introduce the notion of *plastic potential* surface $\psi(\sigma_{ij}) = const.$ The latter is defined in such a way that the outward normal to it specifies the direction of plastic flow. This is illustrated schematically in Figure 1.5, where the plastic strain coordinates are superimposed upon the stress coordinates. The change in the direction of $d\sigma_{ij}$ is said to affect only the magnitude of the plastic strain increment, while the direction of $d\varepsilon_{ij}^p$ remains fully defined by the current stress state σ_{ij}. Noting that the gradient of ψ represents a vector along the outward normal to $\psi(\sigma_{ij}) = const.$, one can write

$$G_{ij} = h\frac{\partial \psi}{\partial \sigma_{ij}}; \quad h = h(\sigma_{ij}, \kappa) \tag{1.26}$$

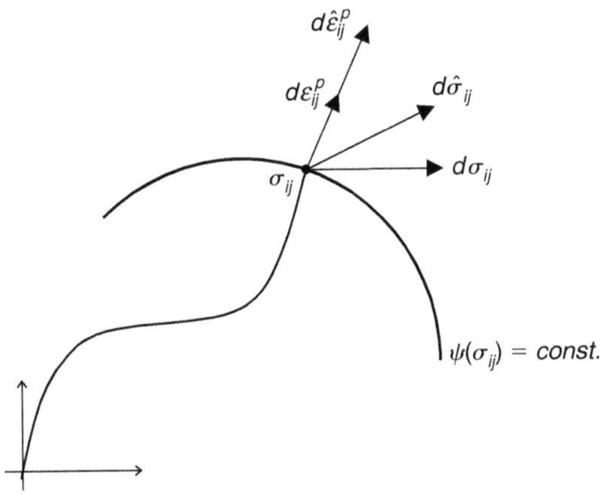

Figure 1.5 The concept of plastic potential and the flow rule

where h, commonly referred to as *hardening function*, is a scalar-valued function of stress and plastic deformation history.

Substituting eq.(1.26) into eq.(1.24) yields

$$d\varepsilon^p_{ij} = h\frac{\partial\psi}{\partial\sigma_{ij}}\,d\bar{f} \tag{1.27}$$

which, given the definition (1.23), results in a constitutive relation in the general form

$$d\varepsilon^p_{ij} = C_{ijkl}\,d\sigma_{kl};\quad C_{ijkl} = h\,\frac{\partial\psi}{\partial\sigma_{ij}}\frac{\partial f}{\partial\sigma_{kl}} \tag{1.28}$$

where C_{ijkl} is the plastic compliance operator.

The representation (1.27) is commonly referred to as a *non-associated flow rule*. In a special case, when the direction of plastic flow is said to remain normal to the loading/failure surface, there is

$$\psi(\sigma_{ij}) = f(\sigma_{ij}) \quad \Rightarrow \quad d\varepsilon^p_{ij} = h\frac{\partial f}{\partial\sigma_{ij}}\,d\bar{f} \tag{1.29}$$

The above form is known as an *associated flow rule* and it is widely adopted in classical metal plasticity [1–3]. It is noted that (1.29) results in the symmetry of the compliance operator (1.28) with respect to the pair of indexes $\{i,j\} \leftrightarrow \{k,l\}$. The associated flow rule has a special significance in the theory of plasticity, since certain variational principles and uniqueness theorems can then be formulated. This aspect is addressed in more detail in the subsequent section.

Before concluding this review, let us recall the basic stress-strain relations in associated flow theory. Assuming the additivity of elastic and plastic strain rates, the response of a *strain-hardening* material can be defined as

$$d\varepsilon_{ij} = \frac{ds_{ij}}{2G} + \frac{1}{9K}\delta_{ij}d\sigma_{kk} + h\frac{\partial f}{\partial \sigma_{ij}}d\bar{f} \quad if \quad \bar{f} = g \wedge d\bar{f} = \frac{\partial f}{\partial \sigma_{ij}}d\sigma_{ij} \geq 0$$

$$d\varepsilon_{ij} = \frac{ds_{ij}}{2G} + \frac{1}{9K}\delta_{ij}d\sigma_{kk} \quad if \quad \bar{f} < g \vee \left(\bar{f} = g \wedge d\bar{f} = \frac{\partial f}{\partial \sigma_{ij}}d\sigma_{ij} \leq 0\right) \tag{1.30}$$

In case of an elastic-*perfectly plastic* material, the flow rule is typically written as

$$d\varepsilon^{p}_{ij} = d\lambda\frac{\partial F}{\partial \sigma_{ij}} \tag{1.31}$$

where $d\lambda \geq 0$ is the so-called *plastic multiplier.* Apparently, there is $d\lambda = h\,d\bar{f}$ for a strain-hardening material. The constitutive relation assumes now the following general form

$$d\varepsilon_{ij} = \frac{ds_{ij}}{2G} + \frac{1}{9K}\delta_{ij}d\sigma_{kk} + d\lambda\frac{\partial F}{\partial \sigma_{ij}} \quad if \quad F = 0 \quad \wedge \quad dF = \frac{\partial F}{\partial \sigma_{ij}}d\sigma_{ij} = 0$$

$$d\varepsilon_{ij} = \frac{ds_{ij}}{2G} + \frac{1}{9K}\delta_{ij}d\sigma_{kk} \quad if \quad F < 0 \quad \vee \quad (F = 0 \wedge dF < 0) \tag{1.32}$$

It is noted that the equations governing the elastoplastic response, as specified in (1.30) and (1.32) above, are not, in general, integrable. In other words, they cannot be reduced to finite relations between stress and strain. This mathematical fact reflects the dependence of the results on the stress history. Thus, the elastoplastic formulation is said to be *path-dependent*. Furthermore, the above constitutive relations do not involve time t. Note, however, that dividing both sides of (1.30)–(1.32) by dt, we can formally pass from increments of strain/stress to strain/stress *rates*, $\dot{\varepsilon}_{ij}, \dot{\sigma}_{ij}$. Then, the equations will resemble those governing the flow of a viscous fluid. This analogy justifies the name of the theory.

1.5 REVIEW OF FUNDAMENTAL POSTULATES OF PLASTICITY; UNIQUENESS OF THE SOLUTION

(i) Condition of irreversibility of deformation

Consider the plastic work per unit volume

$$dW_p = \sigma_{ij}d\varepsilon^{p}_{ij}; \quad W_p = \int \sigma_{ij}d\varepsilon^{p}_{ij} \tag{1.33}$$

Since plastic deformation is an irreversible process, it is essential that the rate (or the increment) of plastic work be non-negative. This statement follows directly from the second law of thermodynamics [11] and implies that *energy dissipated during plastic flow must be positive*, otherwise energy would be generated. For a material with an associated flow rule (1.31), the inequality $dW_p \geq 0$ implies that

$$dW_p = \sigma_{ij} d\varepsilon^p_{ij} = \sigma_{ij} d\lambda \frac{\partial f}{\partial \sigma_{ij}} \geq 0 \tag{1.34}$$

If the loading/failure surface is convex and contains the origin, then

$$\sigma_{ij} \frac{\partial f}{\partial \sigma_{ij}} > 0 \tag{1.35}$$

which results in the requirement that the plastic multiplier must be non-negative, i.e. $d\lambda \geq 0$.

(ii) Postulate of maximum plastic work

Consider a *strain-hardening* material. In the context of a one-dimensional problem, see Figure 1.1d, the notion of hardening implies that an additional loading $d\sigma > 0$ produces an additional strain $d\varepsilon > 0$, so that $(d\sigma)(d\varepsilon) > 0$. Thus, a positive work is done and the material is described as *stable*. The above notion of stability has been generalized by Drucker [12]. Drucker's postulate states that (*i*) during loading and (*ii*) during a complete cycle of additional loading and unloading, the *additional* stress does positive work. This statement is illustrated in Figure 1.6 which shows a schematic representation of a closed cycle of loading-unloading in the stress space. The material is said to be under an initial stress of σ^0_{ij}, point B, which is either on or inside the current loading surface. Consider now the stress cycle BCDB, involving a stress increment $d\sigma_{ij}$ directed towards the exterior of the domain enclosed by the loading surface, and examine the rate of work done by *additional* stress. Since during this cycle the elastic energy is fully recovered, Drucker's definition of stability requires that the rate of plastic work (dW^*_p) be positive. Noting that the path DB is entirely elastic and does not produce any plastic strain, the following inequality is obtained

$$dW^*_p = (\sigma_{ij} - \sigma^0_{ij} + d\sigma_{ij})\ d\varepsilon^p_{ij} > 0 \tag{1.36}$$

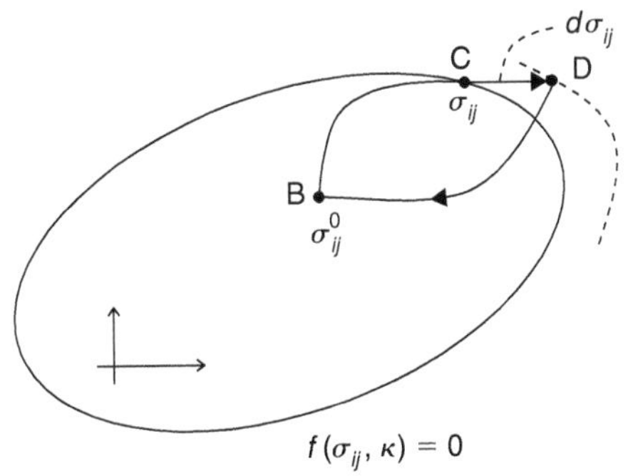

Figure 1.6 A closed stress cycle of loading-unloading

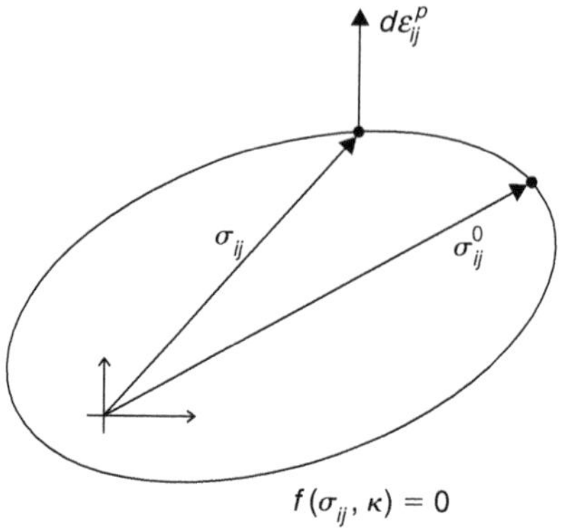

Figure 1.7 Schematic representation of postulate of maximum plastic work

In case when $d\sigma_{ij}$ can be neglected with respect to $(\sigma_{ij} - \sigma_{ij}^0)$ there is

$$(\sigma_{ij} - \sigma_{ij}^0)\, d\varepsilon_{ij}^p > 0 \tag{1.37}$$

which constitutes the *postulate of maximum plastic work*. The interpretation of this postulate is as follows. Assume that σ_{ij}^0 is located on the current loading surface. In this case $\sigma_{ij} d\varepsilon_{ij}^p > \sigma_{ij}^0 d\varepsilon_{ij}^p$, Figure 1.7, which implies that:

The actual rate of energy dissipation is greater than any fictitious one done by an arbitrary state of stress on the same plastic strain rate.

Furthermore, the inequality (1.37) bears two important consequences:

- *Convexity*: The vector $(\sigma_{ij} - \sigma_{ij}^0)$ should always make an acute angle with $d\varepsilon_{ij}^p$ which implies convexity of the loading surface. In other words, if the postulate of maximum plastic work is to be satisfied then the loading surface must be convex. Apparently, for a concave surface, Figure 1.8, there may be $(\sigma_{ij} - \sigma_{ij}^0)\, d\varepsilon_{ij}^p < 0$ which violates (1.37).
- *Normality*: The inequality (1.37) requires the direction of plastic strain increment $d\varepsilon_{ij}^p$ to be normal to the loading surface. In other words, if the postulate of maximum plastic work is to be satisfied then the flow rule must be associated, i.e. the loading and plastic potential surfaces must coincide. This is again schematically illustrated in Figure 1.8; for a non-associated flow there might be $(\sigma_{ij} - \sigma_{ij}^0)\, d\varepsilon_{ij}^p < 0$, which again violates (1.37).

It is noted that the validity of the principle of maximum plastic work also extends to an elastic–perfectly plastic material, thereby requiring the normality rule as well as the convexity of the *failure* surface.

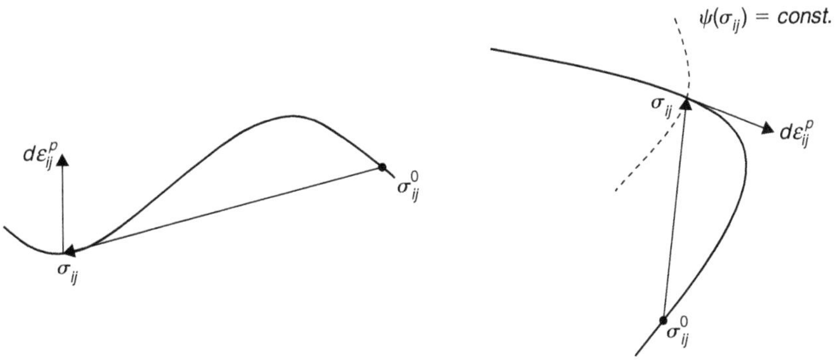

Figure 1.8 Geometric illustration of trajectories violating Drucker's stability postulate

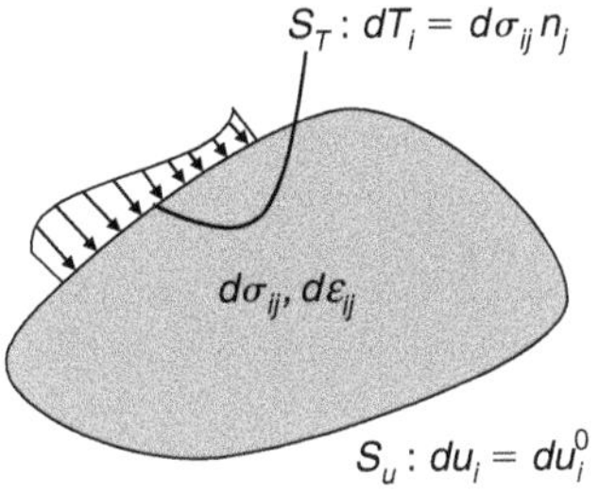

Figure 1.9 A body of volume V bounded by surface S and subjected to traction T_i

Finally, setting $\sigma_{ij} = \sigma_{ij}^0$ in inequality (1.36), i.e. considering only the infinitesimal process, results in

$$d\sigma_{ij}\, d\varepsilon_{ij}^p > 0 \tag{1.38}$$

The above definition is referred to as *stability in small* and it is supplementary to (1.37), which is commonly known as *stability in large*. According to Drucker's definition, a material is considered stable if it satisfies both inequalities, i.e. (1.37) and (1.38). For a stable material one can prove the uniqueness of the solution to a boundary-value problem as well as formulate some variational principles [13], including limit theorems.

(iii) Uniqueness of the solution

Let us consider a hypothetical boundary-value problem as shown schematically in Figure 1.9. At a given instant, the stress and strain fields $\sigma_{ij}, \varepsilon_{ij}$ are known within a body whose volume V is bounded by a closed surface S. Assume now that on S_T, which is a part of S, the surface traction is increased by dT_i, whereas on S_u (i.e. $S - S_T$) the displacement increments du_i are prescribed. In order to prove the uniqueness it must be shown that there is only one stress and strain increment $d\sigma_{ij}, d\varepsilon_{ij}$ within the body which corresponds to an increase in external agencies dT_i, du_i.

Assume that the following conditions are fulfilled:

- Stress increments satisfy the equilibrium and boundary conditions

$$\begin{aligned} d\sigma_{ij,j} &= 0 \quad \text{in V; body forces stationary or absent} \\ d\sigma_{ij}n_j &= dT_i \quad \text{on } S_T \end{aligned} \tag{1.39}$$

- Strain increments can be derived from the displacement field

$$d\varepsilon_{ij} = (du_{i,j} + du_{j,i})/2 \tag{1.40}$$

which satisfies boundary conditions $du_i = du_i^0$ on S_u.

- Stress and strain increments are related by constitutive law

$$\begin{aligned} d\varepsilon_{ij} &= C^e_{ijkl}\, d\sigma_{kl} + C_{ijkl}\, d\sigma_{kl} \\ C^e_{ijkl} &= \frac{1}{2G}\delta_{ik}\delta_{jl} + \frac{2G-3K}{18\,GK}\delta_{ij}\delta_{kl}; \quad C_{ijkl} = h\frac{\partial f}{\partial\sigma_{ij}}\frac{\partial f}{\partial\sigma_{kl}} \end{aligned} \tag{1.41}$$

where $h > 0$ and C^e_{ijkl} and C_{ijkl} are the elastic and plastic compliances, respectively.

Let us admit now that $d\tilde{\sigma}_{ij}, d\sigma_{ij}$ and $d\tilde{\varepsilon}_{ij}, d\varepsilon_{ij}$ are two distinct solutions corresponding to *the same boundary conditions* and consider the integral

$$J = \int_V (d\tilde{\sigma}_{ij} - d\sigma_{ij})(d\tilde{\varepsilon}_{ij} - d\varepsilon_{ij})dV \tag{1.42}$$

Using the kinematic strain-displacement relations and noting that $d\tilde{\sigma}_{ij}$ and $d\sigma_{ij}$ are both symmetric tensors, yields

$$J = \int_V (d\tilde{\sigma}_{ij} - d\sigma_{ij})(d\tilde{u}_i - du_i)_{,j}\, dV \tag{1.43}$$

or

$$J = \int_V [(d\tilde{\sigma}_{ij} - d\sigma_{ij})(d\tilde{u}_i - du_i)]_{,j}\, dV - \int_V (d\tilde{u}_i - du_i)(d\tilde{\sigma}_{ij} - d\sigma_{ij})_{,j}\, dV \tag{1.44}$$

Furthermore, transforming the first integral using Gauss' theorem, one obtains

$$J = \int_S (d\tilde{\sigma}_{ij} - d\sigma_{ij})n_j(d\tilde{u}_i - du_i)dS - \int_V (d\tilde{u}_i - du_i)(d\tilde{\sigma}_{ij} - d\sigma_{ij})_{,j}\, dV \tag{1.45}$$

Now, since both $d\tilde{\sigma}_{ij}$ and $d\sigma_{ij}$ satisfy the equilibrium and static boundary conditions, we finally have

$$J = \int_S (d\tilde{T}_i - dT_i)\,(d\tilde{u}_i - du_i)\,dS \tag{1.46}$$

Since, by the statement of the problem, the boundary conditions are unique on S, i.e. $d\tilde{T}_i = dT_i$ and $d\tilde{u}_i = du_i$, there is $J = 0$; so that, according to (1.42)

$$\int_V (d\tilde{\sigma}_{ij} - d\sigma_{ij})\,(d\tilde{\varepsilon}_{ij} - d\varepsilon_{ij})dV = 0 \tag{1.47}$$

It can be shown now that for a material which is stable in Drucker's sense, the integrand in eq.(1.47) is positive, i.e

$$(d\tilde{\sigma}_{ij} - d\sigma_{ij})(d\tilde{\varepsilon}_{ij} - d\varepsilon_{ij}) > 0 \tag{1.48}$$

Therefore, there must be $d\sigma_{ij} = d\tilde{\sigma}_{ij}$ or $d\varepsilon_{ij} = d\tilde{\varepsilon}_{ij}$ to satisfy (1.47). Since for a strain-hardening material the relation between stress and strain increments (and conversely, between the increments of strain and stress) is unique, the solution to the boundary-value problem, as posed earlier, is also unique. This means that there is only one $d\sigma_{ij}$ and thus $d\varepsilon_{ij}$, which corresponds to an increase in external agencies.

The formal proof of inequality (1.48) can be obtained by invoking the additivity of elastic $d\varepsilon^e_{ij}$ and plastic $d\varepsilon^p_{ij}$ strain increments. In this case, (1.48) can be written in the form

$$(d\tilde{\sigma}_{ij} - d\sigma_{ij})(d\tilde{\varepsilon}^e_{ij} - d\varepsilon^e_{ij}) + (d\tilde{\sigma}_{ij} - d\sigma_{ij})(d\tilde{\varepsilon}^p_{ij} - d\varepsilon^p_{ij}) > 0 \tag{1.49}$$

Given now the constitutive relation (1.41), one has

$$(d\tilde{\sigma}_{ij} - d\sigma_{ij})C^e_{ijkl}\,(d\tilde{\sigma}_{kl} - d\sigma_{kl}) + (d\tilde{\sigma}_{ij} - d\sigma_{ij})C_{ijkl}\,(d\tilde{\sigma}_{kl} - d\sigma_{kl}) > 0 \tag{1.50}$$

Both operators C^e_{ijkl} and C_{ijkl}, as defined in eq.(1.41), are *positive-definite*, which proves that the inequality (1.50) and thus (1.48) remain satisfied.

Finally, it is important to note that the plastic compliance operator C_{ijkl} will loose its positive definiteness in case of a strain softening material and/or when a non-associated flow rule is employed. In this case, the inequality (1.38) will be violated, implying that uniqueness of the solution cannot, in general, be proven.

Chapter 2

Elastic-perfectly plastic formulations

2.1 GENERAL CONSIDERATIONS

In this chapter the simplest idealization, i.e. that of an elastic-perfectly plastic material, is considered. As stated in Section 1.2, the basic hypothesis underlying a phenomenological description of failure is that there exists a *failure function* F of the state of stress σ_{ij} such that, if the ultimate state of the material is reached, the value of this function is constant, say zero,

$$F(\sigma_{ij}) = 0 \tag{2.1}$$

For an isotropic material, F must be isotropic with respect to σ_{ij}, i.e. it must be a function of the basic invariants of σ_{ij}

$$F(\sigma_{ij}) = F(I_1, I_2, I_3) \tag{2.2}$$

or, in fact, any invariant measures which are derived from the basic invariants.

It is noted that in materials with a very low porosity, like metals, the conditions at failure are not affected by the hydrostatic part of σ_{ij}. Therefore,

$$F = F(s_{ij}) = F(J_2, J_3) \tag{2.3}$$

where $J_2 = s_{ij}s_{ij}/2$ and $J_3 = s_{ij}s_{jk}s_{ki}$ are the basic invariants of the stress deviator s_{ij}. On the other hand, for geomaterials, F should include the first invariant I_1. In most of the literature pertaining to plasticity in geomechanics F is assumed as

$$F(\sigma_{ij}) = F(I_1, J_2, J_3) \tag{2.4}$$

Finally, F can also be expressed as a symmetric function of the principal stress magnitudes $\sigma_1, \sigma_2, \sigma_3$. The latter form is particularly attractive as $F(\sigma_1, \sigma_2, \sigma_3) = 0$ can be regarded as a surface in the three-dimensional space where the principal stresses are taken as Cartesian coordinates. All stress trajectories penetrating the domain enclosed by this surface, i.e. satisfying $F < 0$, are admissible. The states corresponding to $F = 0$ define the ultimate stress state, whereas those resulting in $F > 0$ are not permissible.

2.2 GEOMETRIC REPRESENTATION OF THE FAILURE SURFACE

Consider first a failure criterion which is independent of hydrostatic stress, i.e. one of the type $F = F(J_2, J_3) = 0$. If $\sigma_1^0, \sigma_2^0, \sigma_3^0$ is a point on the failure surface, then $\sigma_1 = \sigma_1^0 + I_1/3, \sigma_2 = \sigma_2^0 + I_1/3, \sigma_3 = \sigma_3^0 + I_1/3$ must also lie on the same surface. The latter represent the parametric equations of a line passing through $\sigma_1^0, \sigma_2^0, \sigma_3^0$ and parallel to the line $\sigma_1 = \sigma_2 = \sigma_3$. Thus, it is evident that the surface defined by $F(J_2, J_3) = 0$ is a right cylinder (or prism) with generators perpendicular to any plane $\sigma_1 + \sigma_2 + \sigma_3 = const.$, Figure 2.1. If F includes I_1, the surfaces are no longer cylinders but instead their cross-sections change along the line $\sigma_1 = \sigma_2 = \sigma_3$. This line, which represents the stress space diagonal, is referred to as hydrostatic axis and is defined by a unit normal $\boldsymbol{m} = (1/\sqrt{3})\{1,1,1\}^T$. Furthermore, any plane $\sigma_1 + \sigma_2 + \sigma_3 = const.$, perpendicular to $\boldsymbol{m}$ is called an *octahedral* (π) or *deviatoric* plane.

Referring again to Figure 2.1, let the stress point, represented by a position vector $\mathbf{OP} = \{\sigma_1, \sigma_2, \sigma_3\}^T$, be located on the failure surface. Resolve the stress vector into two components, one along the hydrostatic axis $\mathbf{OO'}$ and the other $\mathbf{O'P}$ perpendicular to it, i.e. confined to the octahedral plane. The vector $\mathbf{OO'}$ along the direction $\boldsymbol{m}$ is defined as

$$\mathbf{OO'} = \frac{1}{\sqrt{3}}(\{\sigma_1, \sigma_2, \sigma_3\}\{1,1,1\}^T)\boldsymbol{m} = \frac{1}{\sqrt{3}} I_1 \boldsymbol{m}; \quad |\mathbf{OO'}| = \frac{1}{\sqrt{3}} I_1 \tag{2.5}$$

so that the vector $\mathbf{O'P}$ in the octahedral plane becomes

$$\mathbf{O'P} = \mathbf{OP} - \mathbf{OO'} = \{s_1, s_2, s_3\}^T, \quad |\mathbf{O'P}| = \sqrt{2}(J_2)^{1/2} \tag{2.6}$$

where s_1, s_2 and s_3 are the principal deviatoric stresses. Thus, any vector in the octahedral plane represents a stress deviator and its magnitude is proportional to the square

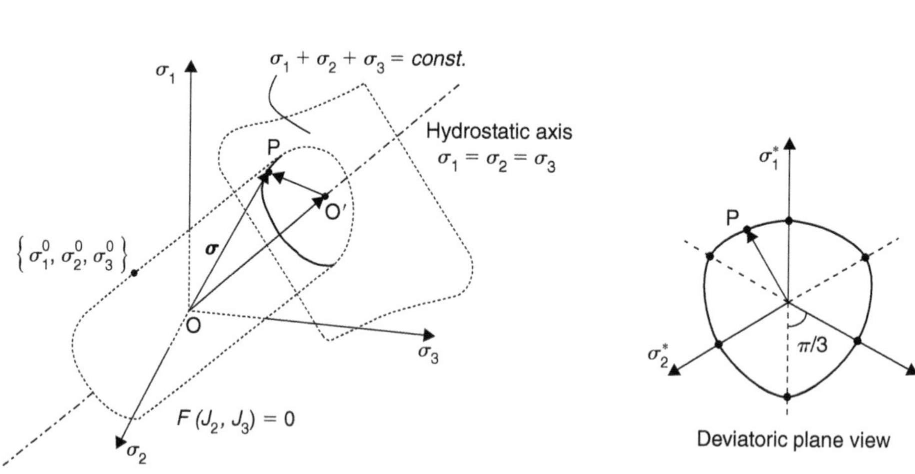

Figure 2.1 Geometric representation of a failure surface $F(J_2, J_3) = 0$ in the principal stress space

root of the second invariant J_2. It is therefore evident that if the function F is of the type $F(I_1, J_2) = 0$ then all cross-sections in the octahedral plane ($I_1 = const.$) are circular ($J_2 = const.$). Such a failure surface is then a surface of revolution about the hydrostatic axis. The simplest form, i.e. $F = F(J_2) = 0$, corresponds to a right circular cylinder.

As demonstrated later in this chapter, the dependence of $F = 0$ on J_3 implies that the octahedral cross-section is represented by some curve, other than a circle. It is desirable (but not necessary, see Section 2.4.3) that this locus be convex with respect to the origin. The isotropy of the material requires that $F = 0$ be a symmetric function of $\sigma_1, \sigma_2, \sigma_3$. Consequently, the failure locus in the octahedral plane must be symmetric about the orthogonal projections of the three principal stress axes $\sigma_1^*, \sigma_2^*, \sigma_3^*$. In other words, the shape of the locus in each of the six segments, indicated in Figure 2.1., must be the same. Therefore, in order to define the whole locus, it is sufficient to consider stress states whose vectors lie in any one of the six segments, corresponding to a chosen ordering of the principal stresses.

2.3 SELECTION OF STRESS INVARIANTS FOR THE MATHEMATICAL DESCRIPTION

The failure function F, which is isotropic in σ_{ij}, may be expressed in terms of any invariant measures derived from the basic invariants. It is convenient to choose invariants which have a direct geometrical interpretation in the principal stress space. Such invariants may be obtained from the basic invariants of the stress deviator s_{ij}.

Consider the characteristic equation of the stress deviator

$$s^3 - J_2 s - J_3 = 0 \tag{2.7}$$

in which s denotes any one of the principal deviatoric stresses. Equation (2.7) may be solved explicitly by substitution [14]

$$s = \frac{2}{\sqrt{3}}(J_2)^{1/2}\sin\theta \tag{2.8}$$

which transforms it into

$$\frac{2}{\sqrt{3}}(J_2)^{3/2}\left(\frac{4}{3}\sin^3\theta - \sin\theta\right) = J_3 \tag{2.9}$$

It can be shown that the expression in brackets is equal to $(-1/3\sin 3\theta)$, so that (2.9) becomes

$$\sin 3\theta = -\frac{3\sqrt{3}}{2}\frac{J_3}{(J_2)^{3/2}} \tag{2.10}$$

Taking 3θ to be in the range $\pm\pi/2$ and noting the periodic nature of $\sin 3\theta = \sin 3(\theta + 2n\pi/3)$, the solution to (2.7), for $s_1 > s_2 > s_3$, is

$$s_1 = \frac{2}{\sqrt{3}}(J_2)^{1/2} \sin\left(\theta + \frac{2}{3}\pi\right); \quad s_2 = \frac{2}{\sqrt{3}}(J_2)^{1/2} \sin\theta;$$
$$s_3 = \frac{2}{\sqrt{3}}(J_2)^{1/2} \sin\left(\theta + \frac{4}{3}\pi\right) \tag{2.11}$$

Thus, the three principal stresses $\sigma_1 > \sigma_2 > \sigma_3$ can be expressed as[1]

$$\begin{Bmatrix} \sigma_1 \\ \sigma_2 \\ \sigma_3 \end{Bmatrix} = \frac{2}{\sqrt{3}}\bar{\sigma} \begin{Bmatrix} \sin\left(\theta + \frac{2}{3}\pi\right) \\ \sin\theta \\ \sin\left(\theta + \frac{4}{3}\pi\right) \end{Bmatrix} - \begin{Bmatrix} \sigma_m \\ \sigma_m \\ \sigma_m \end{Bmatrix} \tag{2.12}$$

where

$$\bar{\sigma} = (J_2)^{1/2}; \quad \sigma_m = -\frac{1}{3}I_1; \quad \theta = \frac{1}{3}\sin^{-1}\left(\frac{-3\sqrt{3}}{2}\frac{J_3}{\bar{\sigma}^3}\right) \wedge -\frac{\pi}{6} \le \theta \le \frac{\pi}{6} \tag{2.13}$$

The set of invariants, $\sigma_m, \bar{\sigma}, \theta$, provides a convenient alternative to the basic invariants for the representation of the failure function F. All these invariants have a direct geometrical interpretation; $\sigma_m, \bar{\sigma}$ through eqs.(2.5) and (2.6), whereas θ is equivalent to Lode's angle [15], i.e. it can be directly identified in the octahedral plane.

The interpretation of θ deserves perhaps some clarification. Referring to Figure 2.2, let the stress point be located at P. According to eq.(2.6), the magnitude of the vector **O′P** equals $\sqrt{2}\bar{\sigma}$. Determine now the projection of **O′P** on the direction of σ_1^*-axis.

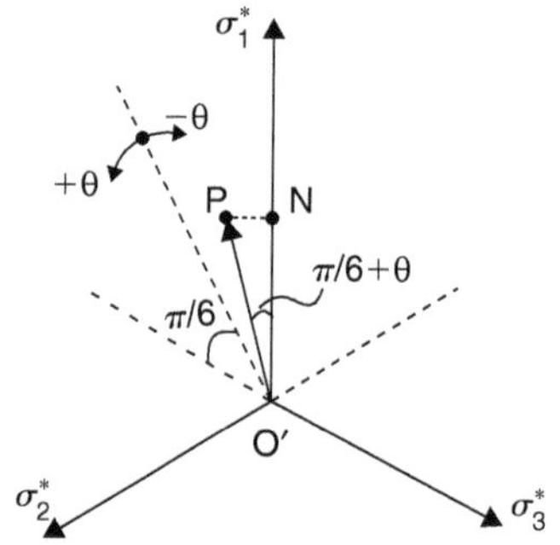

Figure 2.2 Deviatoric plane; definition of angle θ

[1] Standard sign convention of engineering mechanics is used throughout the text, i.e. *tensile* normal stress is considered as *positive*

Since $s_2 = s_3$ along O′N, the unit normal in the direction of σ_1^*-axis is defined by $\{2, -1, -1\}/\sqrt{6}$, so that

$$|\mathrm{O'N}| = \frac{1}{\sqrt{6}}\{s_1, s_2, s_3\}\{2, -1, -1\}^T = \sqrt{\frac{3}{2}}s_1 \tag{2.14}$$

Writing now, according to Figure 2.2,

$$\cos\left(\frac{\pi}{6} + \theta\right) = \frac{|\mathrm{O'N}|}{|\mathrm{O'P}|} = \frac{\sqrt{\frac{3}{2}}s_1}{\sqrt{2}\bar{\sigma}} \tag{2.15}$$

results in

$$s_1 = \frac{2}{\sqrt{3}}\bar{\sigma}\sin\left(\theta + \frac{2}{3}\pi\right) \tag{2.16}$$

which is identical to the first equation in (2.11). The other two equations, for s_2 and s_3, can be verified in a similar manner, thus proving that the above definition of θ is consistent with (2.10).

2.4 FAILURE CRITERIA FOR GEOMATERIALS

2.4.1 Mohr-Coulomb failure criterion

This is the most common criterion employed in the context of geomaterials, in particular soils. In order to define the conditions at failure, consider a plane with normal n_i passing through a representative volume of the material. Let the stress vector, t_i, acting in this plane be resolved into the normal (σ) and shear (τ) components, such that $\sigma = t_i n_i$, $\tau = t_i s_i$, where s_i is a unit vector, normal to n_i, in the plane containing n_i and t_i. The failure across the plane is said to occur if a certain critical combination of τ and σ is reached, i.e.,

$$\tau = f(\sigma) \tag{2.17}$$

The specification of this function typically involves Coulomb's hypothesis, dating from 1773, which postulates a linear relationship between the stress vector components, i.e.,

$$\tau = c - \sigma\tan\phi \tag{2.18}$$

where ϕ and c are termed the *angle of internal friction* and *cohesion*. In order to formulate a general mathematical criterion, equation (2.18) is combined with Mohr circle representation, Figure 2.3. Let all possible states of stress at a point be represented by a family of largest Mohr circles corresponding to, say, $\sigma_1 > \sigma_2 > \sigma_3$. Since the admissible combinations of τ and σ are restricted by $\tau \le c - \sigma\tan\phi$, the failure occurs when the

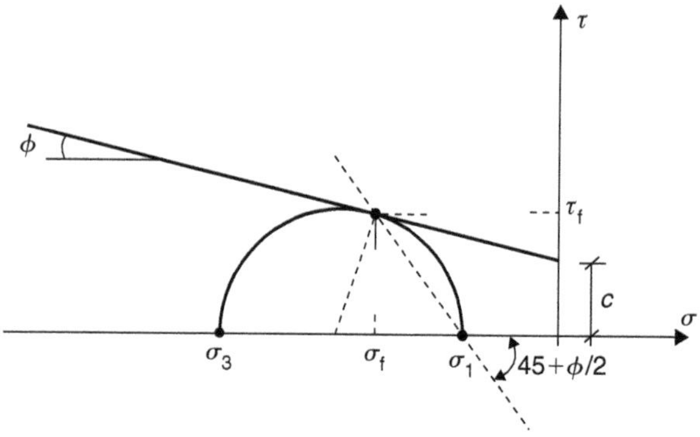

Figure 2.3 Mohr circle defining the conditions at failure

Mohr circle becomes tangential to the Coulomb's envelope (2.18). Thus, according to Figure 2.3, the stress state at failure satisfies

$$\tau = \frac{1}{2}(\sigma_1 - \sigma_3)\cos\phi; \quad \sigma = \frac{1}{2}(\sigma_1 + \sigma_3) + \frac{1}{2}(\sigma_1 - \sigma_3)\sin\phi \tag{2.19}$$

Combining eq.(2.18) with eq.(2.19) yields

$$F = \frac{1}{2}(\sigma_1 - \sigma_3) + \frac{1}{2}(\sigma_1 + \sigma_3)\sin\phi - c\cos\phi = 0 \tag{2.20}$$

which represents the well-established Mohr-Coulomb failure criterion corresponding to the convention $\sigma_1 > \sigma_2 > \sigma_3$. Since the manner in which the principal axes are labeled (viz. 1,2,3) does not affect the form of the failure criterion, eq.(2.20) can be formally re-written for all possible principal stress combinations. In geometrical terms, the set of six equations of the type (2.20) defines an irregular hexagonal pyramid in the three-dimensional principal stress space, Figure 2.4. Its meridians are straight lines whereas the cross-section in the octahedral plane is represented by an irregular hexagon.

It is convenient, in the context of numerical analysis, to express the failure criterion (2.20) in terms of stress invariants $\sigma_m, \overline{\sigma}, \theta$ as defined in Section 2.3. Substitution of eq.(2.13) into eq.(2.20) yields

$$F = \overline{\sigma}(\cos\theta - \sin\theta\sin\phi/\sqrt{3}) - \sigma_m\sin\phi - c\cos\phi = 0 \tag{2.21}$$

By multiplying both sides by a constant $2\sqrt{3}/(3 - \sin\phi)$, eq.(2.21) can be expressed in the form

$$F = \frac{\overline{\sigma}}{g(\theta)} - \eta\sigma_m - \mu = 0 \tag{2.22}$$

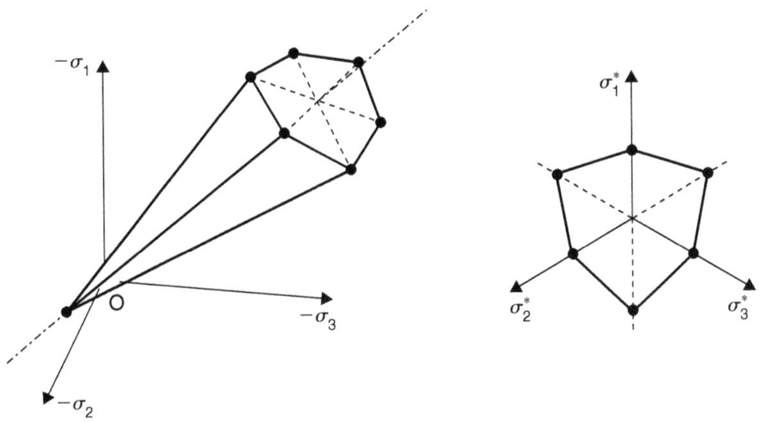

Figure 2.4 Mohr-Coulomb criterion in the principal stress space

where

$$g(\theta) = \frac{3 - \sin\phi}{2\sqrt{3}\cos\theta - 2\sin\theta\sin\phi}; \quad \eta = \frac{2\sqrt{3}\sin\phi}{3 - \sin\phi}; \quad \mu = \frac{2\sqrt{3}c\cos\phi}{3 - \sin\phi} \tag{2.23}$$

and the function $g(\theta)$ satisfies

$$g\left(\frac{\pi}{6}\right) = 1; \quad g\left(-\frac{\pi}{6}\right) = \frac{3 - \sin\varphi}{3 + \sin\varphi} \tag{2.24}$$

It should be noted that the general representation $F = F(\sigma_m, \overline{\sigma}, \theta) = 0$ allows for a very straightforward geometrical interpretation of the failure criteria. Substitution of $\theta = const.$ gives the meridional section relating σ_m and $\overline{\sigma}$, whereas the expression

$$\overline{\sigma} = \overline{\sigma}\Big|_{\theta=\frac{\pi}{6}} g(\theta) \tag{2.25}$$

defines the cross-section in the octahedral plane. The form of $g(\theta)$ as dictated by Mohr-Coulomb criterion, eq.(2.23), results in vertices for $\theta = \pm\pi/6$, Figure 2.4. Note that the conditions associated with $\theta = \pm\pi/6$ are representative of 'triaxial' compression and extension tests, respectively, which are conducted under $\sigma_2 = \sigma_3$. The existence of the vertices presents both conceptual and numerical difficulties since the direction of the gradient tensor is not uniquely defined. Later in this section some approximations to Mohr-Coulomb envelope are discussed based on other continuous and convex functions $g(\theta)$.

The function F, corresponding to Mohr-Coulomb criterion, employs two material constants c and ϕ. Both of these constants should be determined from a minimum of two direct shear tests. Given c and ϕ, the values of η and μ can be calculated from (2.23). Conversely, η and μ can be obtained from a minimum of two 'triaxial' tests, thus providing estimates for c and ϕ from (2.23). In cohesionless materials (e.g. sand) $c = 0$ and, in fact, one direct shear/'triaxial' test is sufficient to identify the function F.

In general though, given the variability of the material, a statistical assessment is always preferable.

2.4.2 Drucker-Prager and other derivative criteria

The Drucker-Prager criterion was formulated in 1952 as a simple extension of von Mises criterion. The functional form is analogous to eq.(2.22) with $g(\theta) = const.$ Taking for instance $g = g(\pi/6) = 1$ gives

$$F = \overline{\sigma} - \eta\sigma_m - \mu = 0 \tag{2.26}$$

Since eq.(2.26) is independent of θ (and thus of the third invariant J_3) the corresponding failure surface is a circular cone. The choice of $g = g(\pi/6) = 1$, with η and μ defined by (2.23), ensures that the cone is passing through the outer corners of the Mohr-Coulomb hexagon (Figure 2.5). Alternatively, assuming $g = g(-\pi/6)$ results in

$$F = \overline{\sigma} - \eta'\sigma_m - \mu' = 0 \tag{2.27}$$

where, according to eqs.(2.22)–(2.24),

$$\eta' = \frac{2\sqrt{3}\sin\phi}{3+\sin\phi}; \quad \mu' = \frac{2\sqrt{3}c\cos\phi}{3+\sin\phi} \tag{2.28}$$

in which case the cone passes through the inner corners of the hexagon. Obviously, a number of other approximations to Mohr-Coulomb envelope may be obtained by an appropriate selection of $g(\theta) = const.$ (for instance, cone inscribed on Mohr-Coulomb hexagon, etc.). Regardless of the selection of constants, (2.26) or (2.27), the

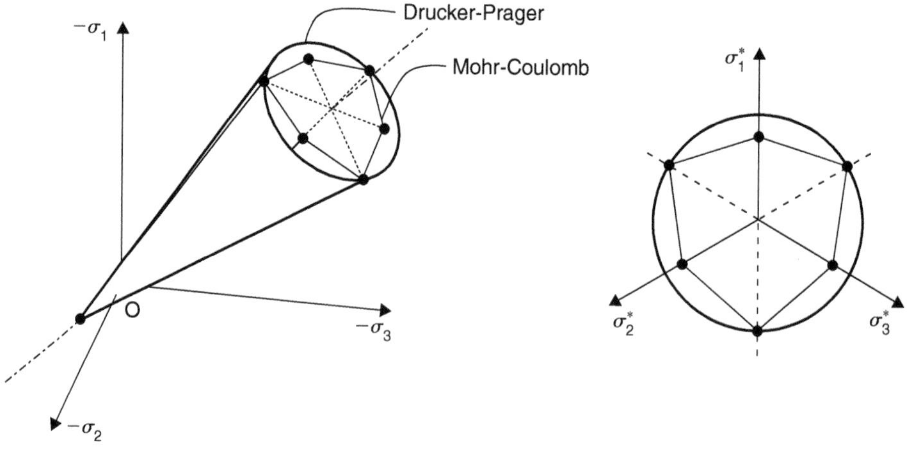

Figure 2.5 Drucker-Prager criterion in the principal stress space

Drucker-Prager criterion is mainly of historic interest since it conflicts with the existing experimental evidence. The latter is clearly in support of representation (2.24) which assumes that for frictional materials the ultimate strength in compression domain ($\theta=\pi/6$) is higher than that in extension ($\theta=-\pi/6$).

It is interesting to note that the typical failure/yield criteria adopted in metal plasticity also represent a particular form of eq.(2.22). The Tresca criterion for instance, corresponds to simply $\phi=0$, i.e.,

$$F=\frac{\overline{\sigma}}{g(\theta)}-\frac{2}{\sqrt{3}}c=0 \tag{2.29}$$

where

$$g(\theta)=\frac{3}{2\sqrt{3}\cos\theta} \tag{2.30}$$

and the constant c, according to eq.(2.18) or eq.(2.20), defines the ultimate stress in pure shear. The criterion (2.29) is independent of the hydrostatic pressure σ_m and in geometrical terms it represents a regular (in view of $g(\theta)=g(-\theta)$) hexagonal prism. The von Mises criterion is obtained by setting $g(\theta)=const.$ in eq.(2.29). The classical representation corresponds to $g(\theta)=g(0)=3/2\sqrt{3}$ which yields

$$F=\overline{\sigma}-c=0 \tag{2.31}$$

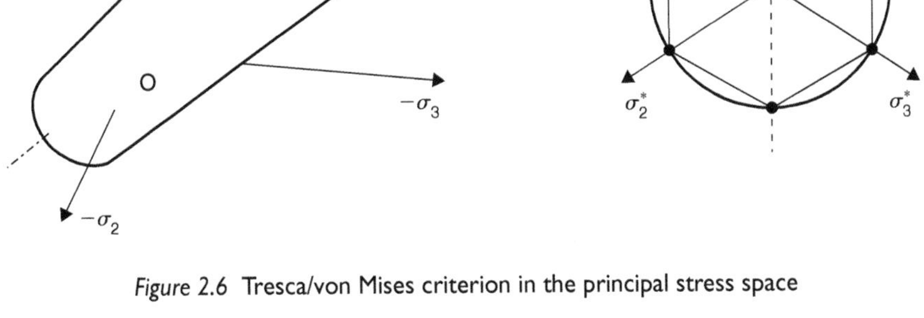

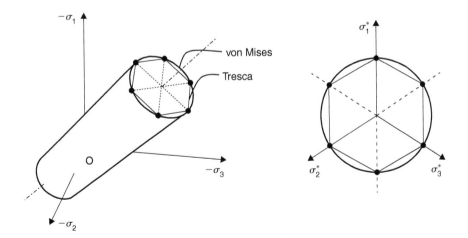

Figure 2.6 Tresca/von Mises criterion in the principal stress space

and represents a cylinder inscribed on Tresca's prism. Once again, various modifications may be considered. Setting for instance $g = g(\pi/6) = 1$, gives

$$F = \bar{\sigma} - \frac{2}{\sqrt{3}} c = 0 \tag{2.32}$$

which corresponds to a cylinder passing through the corners of the prism, Figure 2.6.

2.4.3 Modified criteria based on smooth approximations to Mohr-Coulomb envelope

Consider again a failure criterion in the general form analogous to eq.(2.22), i.e.,

$$F = \frac{\bar{\sigma}}{g(\theta)} - \alpha\sigma_m - \beta = 0 \tag{2.33}$$

where $g(\theta)$ satisfies

$$g\left(\frac{\pi}{6}\right) = 1; \quad g\left(-\frac{\pi}{6}\right) = k \tag{2.34}$$

and α, β, k are material constants. By setting $\alpha = \eta$, $\beta = \mu$ and $k = (3 - \sin\phi)/(3 + \sin\phi)$, as defined in (2.23) and (2.24), various approximations to Mohr-Coulomb envelope can be obtained depending on the choice of $g(\theta)$. It is desirable that $g(\theta)$ be not only a continuous but also a convex function in the entire range $-\pi/6 \leq \theta \leq \pi/6$, satisfying

$$\frac{dg(\theta)}{d\theta} = 0 \quad \text{at} \quad \theta = \pm\frac{\pi}{6} \tag{2.35}$$

The issue of convexity deserves perhaps some clarification. Consider first the postulate of irreversibility (1.34) as discussed in Section 1.5. For an associated flow rule, the rate of energy dissipation is proportional to

$$\sigma_{ij}\frac{\partial F}{\partial \sigma_{ij}} = \sigma_{ij}\left(\frac{\partial F}{\partial \sigma_m}\frac{\partial \sigma_m}{\partial \sigma_{ij}} + \frac{\partial F}{\partial \bar{\sigma}}\frac{\partial \bar{\sigma}}{\partial \sigma_{ij}} + \frac{\partial F}{\partial \theta}\frac{\partial \theta}{\partial \sigma_{ij}}\right) \tag{2.36}$$

It can be shown, using definition (2.13), that

$$\sigma_{ij}\frac{\partial \theta}{\partial \sigma_{ij}} = 0 \tag{2.37}$$

so that, the necessary condition for positive rate of energy dissipation becomes

$$\frac{\partial F}{\partial \sigma_m}\sigma_m + \frac{\partial F}{\partial \bar{\sigma}}\bar{\sigma} \geq 0 \tag{2.38}$$

The inequality (2.38) is independent of θ and thus it is not affected by the shape of the failure surface in the octahedral plane.

The convexity requirement is also linked to Drucker's stability postulate (1.36). In order to prove the uniqueness of the solution to a boundary-value problem, only the stability 'in small' is needed which requires normality but not necessarily convexity. Thus, convexity (in both meridional and octahedral plane) is only required to maintain stability 'in large'. The latter postulate, as stated by Drucker [12], is quite restrictive and may not necessarily apply to all stress trajectories. Therefore, it is quite apparent that the convexity in the octahedral plane should not be regarded as a strict mathematical requirement. It is mainly dictated by the existing experimental evidence. It should be noted that, in order to ensure the convexity of an arbitrary function $g = g(\theta)$ its curvature, ρ,

$$\rho = \left[g^2 + 2\left(\frac{dg}{d\theta}\right)^2 - g\left(\frac{d^2g}{d\theta^2}\right)\right]\left[g^2 + \left(\frac{dg}{d\theta}\right)^2\right]^{-\frac{3}{2}} \tag{2.39}$$

should remain positive [16]. Since the second term in (2.39) is always positive the criterion for convexity reduces to

$$g^2 + 2\left(\frac{dg}{d\theta}\right)^2 - g\left(\frac{d^2g}{d\theta^2}\right) \geq 0 \tag{2.40}$$

Let us now review some known mathematical expressions defining the function $g(\theta)$ which satisfies the constraints (2.34) and (2.35). The simplest is perhaps the expression proposed independently by Gudehus [17] and Argyris et al. [18]

$$g(\theta) = \frac{2k}{(1+k) - (1-k)\sin 3\theta} \tag{2.41}$$

By employing the criterion (2.40) it can be verified that the convexity of (2.41) can only be guaranteed for $k \geq 0.777$, which using the estimate (2.24) corresponds to $\phi \leq 22°$, Figure 2.7.

Another well known expression was proposed by Willam and Warnke [19] as a part of their failure criterion for concrete

$$g(\theta) = \frac{2(1-k^2)\cos\left(\theta + \frac{\pi}{6}\right) + (2k-1)\left[4(1-k^2)\cos^2\left(\theta + \frac{\pi}{6}\right) + 5k^2 - 4k\right]^{1/2}}{4(1-k^2)\cos^2\left(\theta + \frac{\pi}{6}\right) + (2k-1)^2} \tag{2.42}$$

The relation (2.42) is quite complex as it was developed from an elliptical approximation. The latter, however, ensures that (2.42) remains unconditionally convex and smooth in the entire physical range of $0.5 \leq k \leq 1$.

A somewhat simpler expression, which also guarantees convexity in a broad range of k values, was suggested in Ref. [20], i.e.,

$$g(\theta) = \frac{\left(\sqrt{1+a} - \sqrt{1-a}\right)k}{k\sqrt{1+a} - \sqrt{1-a} + (1-k)\sqrt{1-a}\sin 3\theta} \tag{2.43}$$

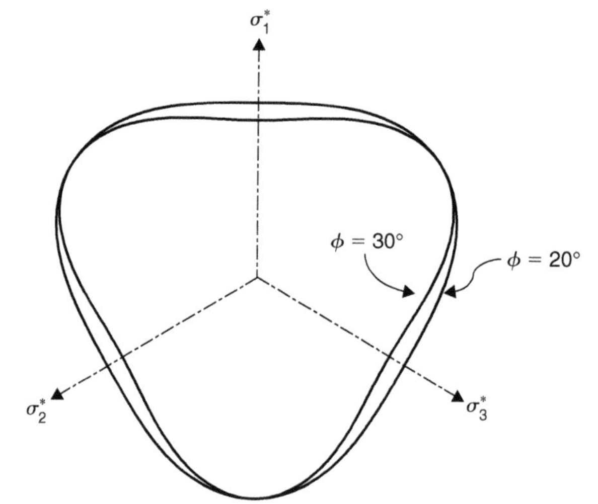

Figure 2.7 Deviatoric cross-sections corresponding to approximation (2.41), for different values of ϕ

where a is a constant and $a \to 1$. (It should be noted that for $a = 1$ the derivative of (2.43) becomes singular at $\theta = \pi/6$). Substituting eq.(2.43) in the criterion (2.40) results, after some algebraic manipulations, in

$$k \geq \frac{7a - 2(1 - \sqrt{1 - a^2})}{9a} \tag{2.44}$$

which for $a = 0.999$, for example, yields $k \geq 0.56$ as the sufficient condition for convexity. Figure 2.8 shows the octahedral cross-sections based on smooth approximations (2.42) and (2.43) to Mohr-Coulomb hexagon. The plots correspond to a selected value of $\phi = 30°$. Other suitable forms of $g(\theta)$ are considered in Ref. [20].

Equation (2.33), combined with one of the representations (2.41) through (2.43), define a failure criterion which is linear in meridional section. It should be noted that for any $g(\theta)$ other than (2.23) the functional form (2.33) depends on the intermediate principal stress, i.e., it is no longer of the type (2.20). Since such a criterion makes no direct reference to Coulomb's law (2.18), the parameters α, β and k may be considered as independent material constants which can be identified from a minimum of three suitably chosen 'triaxial' tests (under $\theta = \pm\pi/6$). For most soils the constant k is typically in the range $0.65 \leq k \leq 0.8$.

2.4.4 Non-linear approximations in meridional sections

Linear form of the function F, eq.(2.33), is adequate for representing the conditions at failure in geomaterials which behave in a stable manner, e.g. normally consolidated and lightly overconsolidated clays, loose and medium dense sands, etc. Certain geomaterials however (e.g. rock materials, concrete) are brittle-plastic, i.e. the failure is

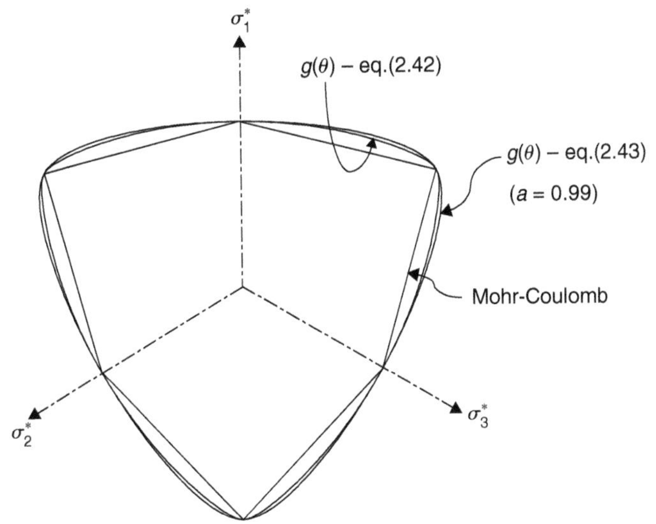

Figure 2.8 Deviatoric cross-sections employing approximations (2.42) and (2.43)

associated with either a ductile (stable) or a brittle (unstable) response, depending on the value of the confining pressure σ_m. In general, the phenomenological description of failure in this class of materials requires the functional form of F which is non-linear in $\overline{\sigma}$ and σ_m. The simplest approximation to F is a quadratic relation

$$F = A_1\left(\frac{\overline{\sigma}}{g(\theta)}\right) + A_2\left(\frac{\overline{\sigma}}{g(\theta)}\right)^2 - (A_3 + \sigma_m) = 0 \tag{2.45}$$

in which A_1 through A_3 are constants. For the purpose of identification, it is convenient to normalize the invariants in (2.45) with respect to any chosen stress measure, so that all the material constants become dimensionless. Introduce, for instance, f_c as the uniaxial compressive strength and define a new set of constants

$$a_1 = A_1, \quad a_2 = A_2 f_c, \quad a_3 = \frac{A_3}{f_c} \tag{2.46}$$

Substituting eq.(2.46) in eq.(2.45) and dividing by f_c yields

$$F = a_1\left(\frac{\overline{\sigma}}{f_c\, g(\theta)}\right) + a_2\left(\frac{\overline{\sigma}}{f_c\, g(\theta)}\right)^2 - \left(a_3 + \frac{\sigma_m}{f_c}\right) = 0 \tag{2.47}$$

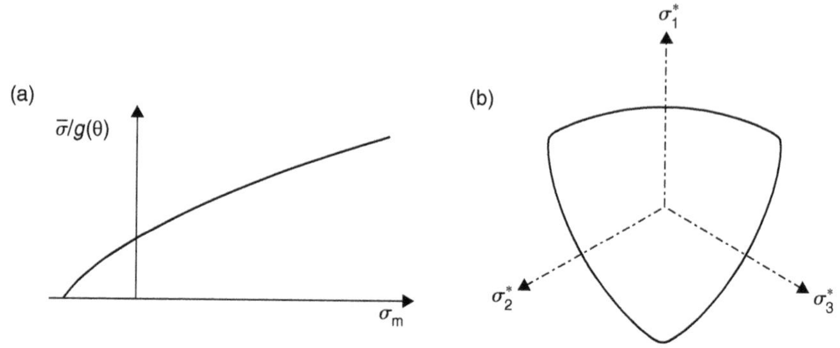

Figure 2.9 Geometric representation of a failure criterion based on the quadratic form (2.45)

Alternatively, solving (2.47) for $\bar{\sigma}/(f_c\, g(\theta))$ results in an equivalent form

$$F = \bar{\sigma} - g(\theta)\bar{\sigma}_c = 0; \;\; \bar{\sigma}_c = \frac{-a_1 + \sqrt{a_1^2 + 4a_2\left(a_3 + \frac{\sigma_m}{f_c}\right)}}{2a_2} f_c \qquad (2.48)$$

in which $\bar{\sigma}_c$ represents the maximum value of $\bar{\sigma}$ in compression ($\theta = \pi/6$) domain.

Figure 2.9 provides a geometrical representation of the above failure criterion. All the meridional sections, corresponding to $\theta = const.$, are segments of a parabola with vertex at $\sigma_m = -a_3 f_c$. The cross-sections in the octahedral plane are described by the function $g(\theta)$, viz. expression (2.25). A suitable form of $g(\theta)$ is either that of (2.42) or (2.43).

The quadratic approximation (2.48) is one of the simplest non-linear forms of F applicable to brittle-plastic materials. The main limitation of (2.48) is perhaps the fact that all the octahedral cross-sections are assumed to be geometrically similar, which conflicts with the existing experimental evidence. In concrete, for example, the parameter k in eq.(2.34) may assume any value within the physical interval $0.5 \le k \le 1$, depending on the value of confining pressure [21]. In other words, the octahedral cross-sections evolve along the hydrostatic axis from a curvilinear triangle ($k \to 0.5$) at low confining pressures to a circle ($k \to 1$) at very high pressures. This effect can be incorporated into (2.48) by regarding g, eq.(2.42) or (2.43), as a composite function $g = g(\theta, k(\sigma_m))$ so that (2.48) becomes

$$F = \bar{\sigma} - g(\theta, k(\sigma_m))\bar{\sigma}_c = 0 \qquad (2.49)$$

The typical failure criteria for concrete [22] are generally of the type (2.49) and differ only in the specification of the function $k(\sigma_m)$. The failure function proposed by William and Warnke [19], for example, has been derived by postulating that the meridional section of (2.49), corresponding to $\theta = -\pi/6$, is a segment of a parabola defined by

$$\bar{\sigma} = k(\sigma_m)\bar{\sigma}_c = \frac{-b_1 + \sqrt{b_1^2 + 4b_2\left(a_3 + \frac{\sigma_m}{f_c}\right)}}{2b_2} f_c \quad for \quad \theta = -\frac{\pi}{6} \tag{2.50}$$

where b_1 and b_2 are constants. Equation (2.50) together with (2.48) correspond to

$$k(\sigma_m) = \frac{a_2\left(-b_1 + \sqrt{b_1^2 + 4b_2\left(a_3 + \frac{\sigma_m}{f_c}\right)}\right)}{b_2\left(-a_1 + \sqrt{a_1^2 + 4a_2\left(a_3 + \frac{\sigma_m}{f_c}\right)}\right)} \tag{2.51}$$

Another possible choice of $k = k(\sigma_m)$ could be an exponential form [21]

$$k = 1 - k_0 \exp\left(-k_1\left(a_3 + \frac{\sigma_m}{f_c}\right)\right) \tag{2.52}$$

in which k_0 and k_1 are constants. According to (2.52), $k \to 1$ for $\sigma_m \to \infty$, whereas for $\sigma_m \to -a_3 f_c$ there is $k \to 1 - k_0$ implying that $k_0 \to 0.5$. In particular, one may choose $g(\theta)$ according to (2.43) with $k_0 \to 0.44$ so that to ensure that the convexity criterion, eq.(2.44), is met.

The failure function (2.49), with g given by (2.42) or (2.43) and $k(\sigma_m)$ defined by (2.51) or (2.52), employs five material constants which can be identified from a series of suitably chosen 'triaxial' tests (under $\theta = \pm\pi/6$). In general, a minimum of five such tests are required, one of them being the uniaxial compression. The identification process can be simplified by making some a priori assumptions consistent with experimental data. Taking, for instance, the strength under hydrostatic tension to be approximately $0.1 f_c$ (see Ref. [23]) and substituting $\sigma_1 = \sigma_2 = \sigma_3 = 0.1 f_c$ in eq.(2.47) or (2.49) results in $a_3 \to 0.1$. Thus, two axial compression tests (under initial confinement $\sigma_m^0 = 0$ and $\sigma_m^0 > 0$) are sufficient to uniquely identify a_1 and a_2, whereas any two lateral compression tests ($\theta = -\pi/6$) may be used to determine b_1 and b_2 from (2.51) or k_0 and k_1 from (2.52).

2.5 DERIVATION OF CONSTITUTIVE RELATION

Given the functional form of $F(\sigma_{ij})$, let us focus our attention now on the specification of the constitutive relation. If the material is regarded as an elastic-perfectly plastic one, then for stress trajectories penetrating the domain enclosed by the failure surface $F = 0$ the response is governed by Hooke's law, i.e,

$$d\sigma_{ij} = D^e_{ijkl}\, d\varepsilon_{kl}, \quad if \quad F < 0 \quad or \quad F = 0 \wedge \frac{\partial F}{\partial \sigma_{ij}} d\sigma_{ij} < 0 \tag{2.53}$$

Here, D^e_{ijkl} is the elastic stiffness tensor, which is the inverse of the compliance operator C^e_{ijkl} as defined earlier in eq.(1.41). In the elastoplastic range, the phenomenological description of the deformation process can be derived from the basic notions of the flow theory. Assume first the additivity of elastic and plastic differential strain increments, i.e.

$$d\varepsilon_{ij} = d\varepsilon^e_{ij} + d\varepsilon^p_{ij} \tag{2.54}$$

The elastic strain increments satisfy Hooke's law, so that

$$d\sigma_{ij} = D^e_{ijkl}(d\varepsilon_{kl} - d\varepsilon^p_{kl}) \tag{2.55}$$

whereas the plastic ones can be derived from a non-associated flow rule (1.27)

$$d\varepsilon^p_{ij} = d\lambda \frac{\partial \psi}{\partial \sigma_{ij}} \tag{2.56}$$

in which $\psi(\sigma_{ij}) = const.$ represents the plastic potential. During the plastic flow the failure surface remains stationary, so that the consistency condition reads

$$dF = \frac{\partial F}{\partial \sigma_{ij}} d\sigma_{ij} = 0 \tag{2.57}$$

Substituting eqs.(2.55) and (2.56) into (2.57) results in

$$dF = \frac{\partial F}{\partial \sigma_{ij}} D^e_{ijkl}\, d\varepsilon_{kl} - \frac{\partial F}{\partial \sigma_{ij}} D^e_{ijkl} \frac{\partial \psi}{\partial \sigma_{kl}} d\lambda = 0 \tag{2.58}$$

from which

$$d\lambda = \frac{1}{H}\left(\frac{\partial F}{\partial \sigma_{ij}} D^e_{ijkl}\, d\varepsilon_{kl}\right), \quad H = \frac{\partial F}{\partial \sigma_{pq}} D^e_{pqrs} \frac{\partial \psi}{\partial \sigma_{rs}} \tag{2.59}$$

Thus, the plastic strain increments can be uniquely determined from the flow rule (2.56) in which $d\lambda$ is a positive scalar defined by eq.(2.59). Finally, the incremental stress-strain relation can now be derived by substituting eqs.(2.56) and (2.59) in eq.(2.55)

$$d\sigma_{ij} = D^{ep}_{ijkl}\, d\varepsilon_{kl}, \quad if \quad F = 0 \wedge \frac{\partial F}{\partial \sigma_{ij}} d\sigma_{ij} = 0 \tag{2.60}$$

where

$$D^{ep}_{ijkl} = D^e_{ijkl} - \frac{1}{H}\left(D^e_{ijpq} \frac{\partial \psi}{\partial \sigma_{pq}} \frac{\partial F}{\partial \sigma_{rs}} D^e_{rskl}\right) \tag{2.61}$$

Equation (2.61) defines explicitly the constitutive tensor for an elastic-perfectly plastic material. Given a prescribed strain increment $d\varepsilon_{ij}$, the corresponding stress increment can be uniquely determined. The converse however is not true, i.e. if an arbitrary stress increment satisfying the loading criterion is prescribed, the elastic strain increment is known, but $d\lambda$ and thus the magnitude of the plastic strain increment are indeterminate from (2.56).

The components of the constitutive tensor (2.61) are function of the gradient tensors which, in turn, depend on the current state of stress σ_{ij}. If $F=0$ is expressed as $F=F(\sigma_m, \overline{\sigma}, \theta)=0$, then

$$\frac{\partial F}{\partial \sigma_{ij}} = \frac{\partial F}{\partial \sigma_m}\frac{\partial \sigma_m}{\partial \sigma_{ij}} + \frac{\partial F}{\partial \overline{\sigma}}\frac{\partial \overline{\sigma}}{\partial \sigma_{ij}} + \frac{\partial F}{\partial \theta}\frac{\partial \theta}{\partial \sigma_{ij}} \tag{2.62}$$

Differentiating θ, eq.(2.13), with respect to σ_{ij}, one obtains

$$\frac{\partial \theta}{\partial \sigma_{ij}} = \frac{\partial \theta}{\partial \overline{\sigma}}\frac{\partial \overline{\sigma}}{\partial \sigma_{ij}} + \frac{\partial \theta}{\partial J_3}\frac{\partial J_3}{\partial \sigma_{ij}} = \frac{\sqrt{3}}{2\overline{\sigma}^3 \cos 3\theta}\left(\frac{3J_3}{\overline{\sigma}}\frac{\partial \overline{\sigma}}{\partial \sigma_{ij}} - \frac{\partial J_3}{\partial \sigma_{ij}}\right) \tag{2.63}$$

Thus, eq.(2.62) can be written as

$$\frac{\partial F}{\partial \sigma_{ij}} = c_1 \frac{\partial \sigma_m}{\partial \sigma_{ij}} + c_2 \frac{\partial \overline{\sigma}}{\partial \sigma_{ij}} + c_3 \frac{\partial J_3}{\partial \sigma_{ij}} \tag{2.64}$$

where

$$c_1 = \frac{\partial F}{\partial \sigma_m}, \quad c_2 = \frac{\partial F}{\partial \overline{\sigma}} - \frac{\tan 3\theta}{\overline{\sigma}}\frac{\partial F}{\partial \theta}, \quad c_3 = \frac{-3\sqrt{3}}{2\overline{\sigma}^3 \cos 3\theta}\frac{\partial F}{\partial \theta} \tag{2.65}$$

and

$$\frac{\partial \sigma_m}{\partial \sigma_{ij}} = -\frac{1}{3}\delta_{ij}, \quad \frac{\partial \overline{\sigma}}{\partial \sigma_{ij}} = \frac{1}{2\overline{\sigma}} s_{ij}, \quad \frac{\partial J_3}{\partial \sigma_{ij}} = s_{ik}s_{kj} - \frac{2}{3}\overline{\sigma}^2 \delta_{ij} \tag{2.66}$$

2.5.1 Matrix formulation

The constitutive relation derived in the previous section can be conveniently expressed in the matrix form, suitable for use in the numerical analysis. Define first the vector forms of differential stress and strain increments as

$$\begin{aligned} d\boldsymbol{\sigma} &= \{d\sigma_x, d\sigma_y, d\sigma_z, d\sigma_{xy}, d\sigma_{yz}, d\sigma_{zx}\}^T, \\ d\boldsymbol{\varepsilon} &= \{d\varepsilon_x, d\varepsilon_y, d\varepsilon_z, d\gamma_{xy}, d\gamma_{yz}, d\gamma_{zx}\}^T \end{aligned} \tag{2.67}$$

where γ's represent the engineering shear strain, i.e. $\gamma_{xy}=2\varepsilon_{xy}$, etc.
Let $d\boldsymbol{\varepsilon} = d\boldsymbol{\varepsilon}^e + d\boldsymbol{\varepsilon}^p$, where

$$d\boldsymbol{\varepsilon}^p = d\lambda \frac{\partial \psi}{\partial \boldsymbol{\sigma}} \quad \mathit{for} \quad F=0 \wedge dF = \left(\frac{\partial F}{\partial \boldsymbol{\sigma}}\right)^T d\boldsymbol{\sigma} = 0 \tag{2.68}$$

In equation (2.68) the gradient of the function F is expressed as

$$\frac{\partial F}{\partial \boldsymbol{\sigma}} = \left\{\frac{\partial F}{\partial \sigma_x}, \frac{\partial F}{\partial \sigma_y}, \frac{\partial F}{\partial \sigma_z}, 2\frac{\partial F}{\partial \sigma_{xy}}, 2\frac{\partial F}{\partial \sigma_{yz}}, 2\frac{\partial F}{\partial \sigma_{zx}}\right\}^T \tag{2.69}$$

with a similar definition holding for the potential function ψ. The representation (2.69) is preserved only if $F = F(\sigma_{ij})$ is expressed as a function of all nine stress components. If the symmetry of σ_{ij} is imposed in F (i.e. $\sigma_{xy} = \sigma_{yx}$ ….., etc.) the factor of 2, accompanying the shear terms, is omitted so that (2.68) is consistent with (2.56) and (2.57).

The constitutive matrix can be derived following the procedure analogous to that adopted before. Writing eq.(2.55) as

$$d\boldsymbol{\sigma} = [D^e](d\boldsymbol{\varepsilon} - d\boldsymbol{\varepsilon}^p) = [D^e]\left(d\boldsymbol{\varepsilon} - d\lambda\frac{\partial \psi}{\partial \boldsymbol{\sigma}}\right) \tag{2.70}$$

where $[D^e]$ is the elastic constitutive matrix, and using the consistency condition, $dF = 0$ in eq.(2.68), leads to

$$d\lambda = \frac{1}{H}\left(\left(\frac{\partial F}{\partial \boldsymbol{\sigma}}\right)^T [D^e] d\boldsymbol{\varepsilon}\right), \quad H = \left(\frac{\partial F}{\partial \boldsymbol{\sigma}}\right)^T [D^e]\frac{\partial \psi}{\partial \boldsymbol{\sigma}} \tag{2.71}$$

Substituting (2.71) into (2.70) and rearranging, results in

$$d\boldsymbol{\sigma} = [D^{ep}]d\boldsymbol{\varepsilon}, \quad [D^{ep}] = [D^e] - \frac{1}{H}\left([D^e]\frac{\partial \psi}{\partial \boldsymbol{\sigma}}\left(\frac{\partial F}{\partial \boldsymbol{\sigma}}\right)^T [D^e]\right) \tag{2.72}$$

which is equivalent to (2.60) and (2.61). As mentioned before, the matrix $[D^{ep}]$ is singular, implying that there is no unique response in strain increment for a given stress increment. Moreover, $[D^{ep}]$ is, in general, non-symmetric. The symmetry is recovered only if $F = \psi$, i.e. for an associated flow rule.

Finally, the gradient of F, eq.(2.64), can be expressed as

$$\frac{\partial F}{\partial \boldsymbol{\sigma}} = c_1\frac{\partial \sigma_m}{\partial \boldsymbol{\sigma}} + c_2\frac{\partial \bar{\sigma}}{\partial \boldsymbol{\sigma}} + c_3\frac{\partial J_3}{\partial \boldsymbol{\sigma}} \tag{2.73}$$

where, according to eqs.(2.66) and (2.69)

$$\begin{aligned}
\frac{\partial \sigma_m}{\partial \boldsymbol{\sigma}} &= -\frac{1}{3}\{1,1,1,0,0,0\}^T; \quad \frac{\partial \bar{\sigma}}{\partial \boldsymbol{\sigma}} = -\frac{1}{2\bar{\sigma}}\{s_x, s_y, s_z, 2\sigma_{xy}, 2\sigma_{yz}, 2\sigma_{zx}\}^T \\
\frac{\partial J_3}{\partial \boldsymbol{\sigma}} &= \Big\{(s_y s_z - \sigma_{yz}^2) + \frac{1}{3}\bar{\sigma}^2, (s_x s_z - \sigma_{xz}^2) + \frac{1}{3}\bar{\sigma}^2, (s_x s_y - \sigma_{xy}^2) + \frac{1}{3}\bar{\sigma}^2, \\
&\quad 2(\sigma_{yz}\sigma_{xz} - s_z\sigma_{xy}), 2(\sigma_{xz}\sigma_{xy} - s_x\sigma_{yz}), 2(\sigma_{xy}\sigma_{yz} - s_y\sigma_{xz})\Big\}^T
\end{aligned} \tag{2.74}$$

The coefficients c_1, c_2 and c_3 are defined by eq.(2.65) and can easily be determined for any specific form of the function $F(\boldsymbol{\sigma})$.

2.6 CONSEQUENCES OF A NON-ASSOCIATED FLOW RULE

The elastic-perfectly plastic idealization is typically used in the context of geomaterials for which the conditions at failure are described by either Mohr-Coulomb criterion or any smooth approximation to it. For this class of materials the applicability of an associated flow rule is often questioned. The main argument here is that the linear form (2.33) results in

$$d\varepsilon^p_{ii} = d\lambda \frac{\partial F}{\partial \sigma_{ii}} = d\lambda \alpha > 0 \tag{2.75}$$

i.e., the plastic deformation is accompanied by a continuing dilation (increase in volume). This conflicts with the existing experimental evidence which indicates that once the ultimate state in a homogeneous material is reached, the volume changes become negligible. The last observation prompted the researchers to use perfectly-plastic formulations incorporating a non-associated flow rule, viz. eq.(2.56), which requires specification of an appropriate form of the plastic potential.

The plastic potential function ψ is supposed to be a scalar-valued function of σ_{ij} alone, i.e. it is independent of the deformation history. For an isotropic material, ψ is subjected to similar requirements as those imposed on the function F. Thus, ψ should be taken as an isotropic function of σ_{ij}, symmetric in σ_1, σ_2 and σ_3. In particular, it is convenient to adopt the representation analogous to that for F, i.e. $\psi = \psi(\sigma_m, \overline{\sigma}, \theta)$. Furthermore, the condition of irreversibility following from the second law of thermodynamics demands that $\sigma_{ij} d\varepsilon^p_{ij} \geq 0$, which for a non-associated flow rule (2.56) requires

$$\sigma_{ij} \frac{\partial \psi}{\partial \sigma_{ij}} \geq 0 \tag{2.76}$$

The above inequality is satisfied for all plastic potential surfaces $\psi = const.$ which are convex and contain the origin of the stress space. Once again, the convexity in the octahedral plane is not strictly required in view of (2.37).

An admissible form of $\psi = const.$, often used in combination with (2.33), is

$$\psi = \frac{\overline{\sigma}}{g(\theta)} - \overline{\alpha}\sigma_m = const. \tag{2.77}$$

where $\overline{\alpha}$ is a constant. Usually, $\overline{\alpha}$ is expressed as (see eq.(2.23))

$$\overline{\alpha} = \frac{2\sqrt{3}\sin\varphi}{3 - \sin\varphi} \tag{2.78}$$

where φ is referred to as the dilatancy angle. According to eq.(2.77)

$$d\varepsilon^p_{ii} = d\lambda\overline{\alpha} > 0 \tag{2.79}$$

and $\overline{\alpha} < \alpha$ (i.e. $\varphi < \phi$), which is consistent with the experimental evidence.

The question of whether or not to employ a non-associated flow rule is certainly debatable. The main arguments used for it, is that the predicted deformation field is more accurate. This may not generally be true in view of a rather drastic perfectly-plastic idealization. It should also be remembered that the use of a non-associated law has some important implications. First of all, it leads to a non-symmetric constitutive matrix, eq.(2.72). Thus, in the context of a finite element approach, the resulting tangential stiffness matrices also remain non-symmetric requiring the use of more complex solution routines. Furthermore, one cannot, in general, ensure the uniqueness of the solution to a boundary-value problem (see Section 1.5).

Chapter 3

Isotropic strain-hardening formulations

The elastic – perfectly plastic idealization, as discussed in the previous section, is rather simplistic and it is not capable of describing a number of distinct features in the mechanical response of porous media. This is particularly evident when examining the complexity of characteristics of geomaterials which are saturated with fluid. A much more realistic representation can be obtained by invoking the notion of strain-hardening. In this chapter several basic approaches incorporating isotropic strain-hardening will be examined and their numerical performance evaluated.

3.1 'TRIAXIAL' TESTS AND THEIR MATHEMATICAL REPRESENTATION

The most common experimental tests for geomaterials, in particular those naturally occurring such as soils, are the 'triaxial' tests. The tests involve a cylindrical specimen which is enclosed inside a thin rubber membrane and placed in a plastic chamber filled with water. The specimen can be tested at different initial confining stress by application of pressure to the fluid; the failure of the sample is typically triggered by applying additional vertical stress. The tests normally involve a saturated sample and can be either *drained* or *undrained*. In the former case, the excess of pore pressure is equal to zero, which implies that the conditions are the same, in mechanical terms, as in a dry sample. In undrained tests the drainage valve is kept closed, so that the water cannot escape from the sample resulting in generation of excess pore pressure.

Referring to Figure 3.1, the 'triaxial' tests are conducted in the principal stress configuration; the vertical stress is typically labelled as σ_1, while $\sigma_2 = \sigma_3$. In order to provide the mathematical representation of mechanical response, it is convenient to introduce the stress invariants $\{p, q\}$ defined as

$$p = -\frac{1}{3}(\sigma_1 + 2\sigma_3); \quad q = \sigma_3 - \sigma_1 \tag{3.1}$$

Here, p is the mean normal stress, while q specifies the deviatoric stress intensity. Both stress parameters are functions of the basic invariants of the stress tensor/deviator. In particular, invoking the definition (2.13) under $\sigma_2 = \sigma_3$, yields

$$\sigma_m = -\frac{1}{3}(\sigma_1 + 2\sigma_3) = p; \quad \bar{\sigma} = \frac{1}{\sqrt{3}}|\sigma_3 - \sigma_1| = \frac{|q|}{\sqrt{3}} \tag{3.2}$$

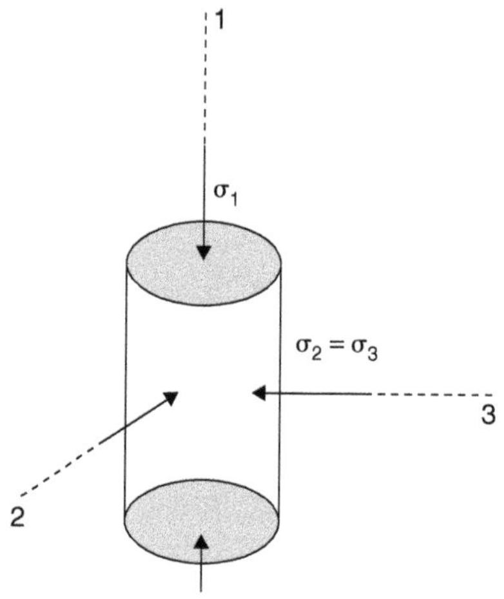

Figure 3.1 'Triaxial' configuration

and also

$$J_3 = -\frac{2}{27}(\sigma_3 - \sigma_1)^3 \quad \Rightarrow \quad \theta = \pm\frac{\pi}{6} \tag{3.3}$$

The strain rate invariants which are compatible with (3.1) are defined as

$$d\varepsilon_v = -(d\varepsilon_1 + 2d\varepsilon_3); \quad d\varepsilon_q = \frac{2}{3}(d\varepsilon_3 - d\varepsilon_1) \tag{3.4}$$

so that the rate of work becomes

$$dW = p\,d\varepsilon_v + q\,d\varepsilon_q = \sigma_1 d\varepsilon_1 + 2\sigma_3\,d\varepsilon_3 \tag{3.5}$$

It is noted that ε_v is the volumetric strain whereas ε_q, i.e. the deviatoric strain, is a measure of the distortion of the sample.

3.1.1 Mohr-Coulomb criterion in 'triaxial' space

Given the representation (3.1), consider now the predictive abilities of an elastic – perfectly plastic formulation. For this purpose, let us examine first the conditions at failure. In order to specify the form of the Mohr-Coulomb criterion in the 'triaxial' configuration, write eq.(3.1) in the inverse form, i.e.

$$\sigma_1 = -p - \frac{2}{3}q; \quad \sigma_3 = -p + \frac{1}{3}q \tag{3.6}$$

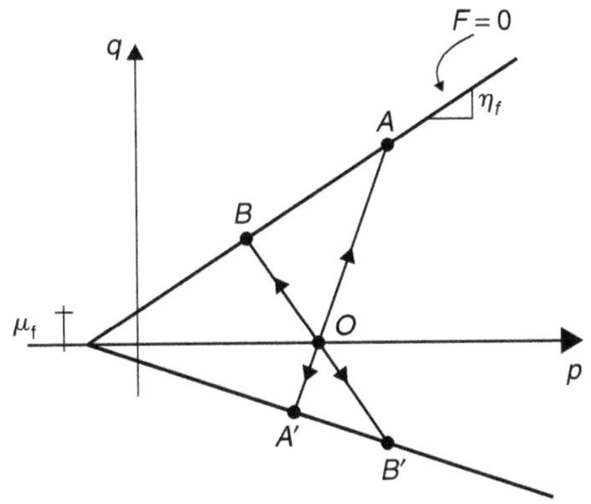

Figure 3.2 Mohr-Coulomb failure envelope and typical stress trajectories in $\{p, q\}$ space

Noting that for $\sigma_3 > \sigma_1$ (i.e. $q > 0$), the Mohr-Coulomb criterion (2.20) takes the form

$$F = \frac{1}{2}(\sigma_3 - \sigma_1) + \frac{1}{2}(\sigma_3 + \sigma_1)\sin\phi - c\cos\phi = 0 \tag{3.7}$$

and substituting the set of eqs.(3.6) results in

$$F = q - \eta_f p - \mu_f = 0 \quad (for\ q > 0) \tag{3.8}$$

where

$$\eta_f = \frac{6\sin\phi}{3 - \sin\phi}; \quad \mu_f = \frac{6c\cos\phi}{3 - \sin\phi} \tag{3.9}$$

Similarly, in the extension domain (i.e. $q < 0$)

$$F = q + \tilde{\eta}_f p + \tilde{\mu}_f = 0 \quad (for\ q < 0) \tag{3.10}$$

where

$$\tilde{\eta}_f = \frac{6\sin\phi}{3 + \sin\phi}; \quad \tilde{\mu}_f = \frac{6c\cos\phi}{3 + \sin\phi} \tag{3.11}$$

Thus, in the 'triaxial' space $\{p, q\}$, the conditions at failure are defined by linear forms (3.8) and (3.10), Figure 3.2. Note that p-axis is not an axis of symmetry, implying that the strength in extension domain ($q < 0$) is generally lower than that in compression domain ($q > 0$).

Given the representation above, typical stress trajectories associated with 'triaxial' tests can now be examined. The most common one is the *axial compression*. The test involves subjecting the sample to an initial cell pressure (point O in Figure 3.2) which is followed by application of additional compressive stress $d\sigma_1 < 0$ under $d\sigma_2 = d\sigma_3 = 0$. Note again that the indexes 1-3 refer here to the geometry of the sample (Figure 3.1), i.e. σ_1 is identified with the vertical stress component, while σ_2 and σ_3 define the lateral components. Given eq.(3.1), the stress trajectory for this test corresponds to OA, Figure 3.2, and satisfies $dq/dp = 3$ while $dq > 0$. Similarly, for *axial extension* there is $d\sigma_1 > 0$, so that $dq/dp = 3$ while $dq < 0$ (trajectory OA'). *Lateral compression* corresponds to $d\sigma_1 = 0$ while $d\sigma_2 = d\sigma_3 < 0$ and results in the stress path OB', which satisfies $dq/dp = -3/2$ and $dq < 0$. Apparently, the *lateral extension* results in trajectory OB which has the same slope of -1.5. The stress coordinates at A, A', B and B' satisfy $F = 0$, implying that the ultimate state of the material is reached.

3.1.2 On the behaviour of a perfectly plastic Mohr-Coulomb material

Consider now the material as elastic–perfectly plastic and examine the governing constitutive relation. Under $\sigma_2 = \sigma_3$, the elastic response can be described by

$$\begin{Bmatrix} dp \\ dq \end{Bmatrix} = \begin{bmatrix} K & 0 \\ 0 & 3G \end{bmatrix} \begin{Bmatrix} d\varepsilon_v \\ d\varepsilon_q \end{Bmatrix} \tag{3.12}$$

where K and G are the bulk and shear moduli, respectively. The elastoplastic operator $[D^{ep}]$ is defined according to eq.(2.72). Assuming an associated flow rule and taking $F = F(p, q)$ in accordance with eq.(3.8), results in

$$[D^{ep}] = \begin{bmatrix} K & 0 \\ 0 & 3G \end{bmatrix} - \frac{1}{K\eta_f^2 + 3G} \begin{bmatrix} K^2\eta_f^2 & -3GK\eta_f \\ -3GK\eta_f & 9G^2 \end{bmatrix} \tag{3.13}$$

so that

$$\begin{Bmatrix} dp \\ dq \end{Bmatrix} = \begin{bmatrix} \left(K - \dfrac{K^2\eta_f^2}{K\eta_f^2 + 3G}\right) & \left(\dfrac{3GK\eta_f}{K\eta_f^2 + 3G}\right) \\ \left(\dfrac{3GK\eta_f}{K\eta_f^2 + 3G}\right) & \left(3G - \dfrac{9G^2}{K\eta_f^2 + 3G}\right) \end{bmatrix} \begin{Bmatrix} d\varepsilon_v \\ d\varepsilon_q \end{Bmatrix} \tag{3.14}$$

Note that $[D^{ep}]$, as defined above, is symmetric, which stems from employing an associated flow rule. Also,

$$\det[D^{ep}] = \left(K - \frac{K^2\eta_f^2}{K\eta_f^2 + 3G}\right)\left(3G - \frac{9G^2}{K\eta_f^2 + 3G}\right) - \frac{9G^2K^2\eta_f^2}{(K\eta_f^2 + 3G)^2} = \ldots\ldots = 0 \tag{3.15}$$

so that $[D^{ep}]$ is singular, i.e. there is no unique response in strain rate for an arbitrary stress rate.

Let us examine briefly the basic trends in the mechanical response, as implied by the constitutive relation (3.14). For all *drained tests*, as depicted in Figure 3.2, the response is said to be elastic until the stress trajectory reaches the failure envelope $F=0$. At this stage, the stress state remains stationary and an unlimited plastic flow commences. Clearly, such an idealization is adequate in terms of predicting the conditions at failure; it is not, however, very accurate in terms of describing the strain history.

Consider now an *undrained* response. If the sample is fully saturated and the fluid cannot escape from voids then the overall macroscopic deformation must be consistent with that of the constituents. Let us focus the attention on a class of uncemented granular materials, like soils, for which the Terzaghi's *effective stress principle* applies. The latter takes the form

$$\hat{p}=p+p_w; \quad \hat{q}=q \tag{3.16}$$

so that $p_w=\hat{p}-p$, where $\hat{p}$ is the *total* pressure while p is interpreted as the *effective* pressure.[1] Thus, for saturated soil, the formulation of the problem, viz. eqs.(3.8)–(3.14), is interpreted as being given in terms of *effective* pressure p. Evidently, if no excess of pore pressure is generated, i.e. the conditions are drained, there is $p=\hat{p}$; $q=\hat{q}$.

For most soils, the deformability of grains forming the soil skeleton can be neglected and fluid itself (typically water) may be considered as incompressible as compared to compressibility of the skeleton. In this case, the kinematic constraint of undrained deformation becomes $d\varepsilon_v \to 0$, i.e. no volume change. Let us examine now the *effective* stress trajectory that corresponds to this constraint. In the elastic range, according to eq.(3.12), there is

$$d\varepsilon_v \to 0 \quad \Rightarrow \quad dp \to 0 \quad \Rightarrow \quad \frac{dq}{dp} \to \infty \tag{3.17}$$

Thus, the effective stress path is a vertical line passing through p_0, which represents the initial confining pressure, Figure 3.3a. The above trajectory is valid as long as $F<0$. When the plastic flow commences, the constraint $d\varepsilon_v \to 0$ results in (see eq.(3.14))

$$\frac{dq}{dp}=\frac{D_{22}^{ep}}{D_{12}^{ep}}=\frac{3G(K\eta_f^2+3G)-9G^2}{3GK\eta_f}=\eta_f \tag{3.18}$$

Thus, the effective stress paths moves now along the failure envelope. It is noted that, at any stage of the deformation process, the excess of pore pressure, p_w, can be assessed based on the total stress trajectory by employing the Terzaghi's principle (3.16).

[1] Note that the notation is different here from that in classical Soil Mechanics; this is in order to avoid the use of multiple superscripts later in the presentation

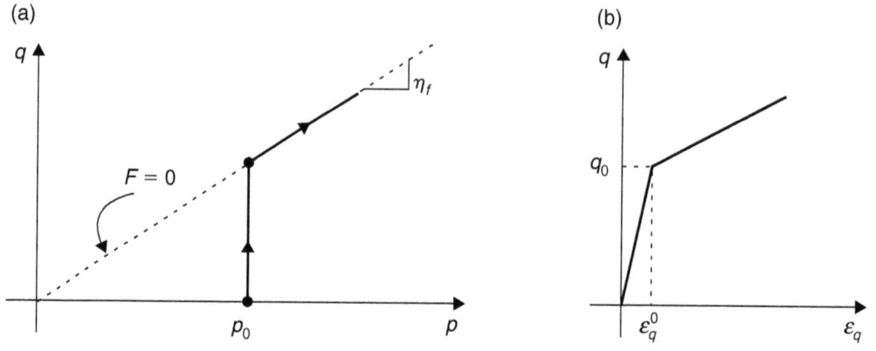

Figure 3.3 Undrained response of an elastic-perfectly plastic granular material

Finally, the stress-strain behaviour corresponding to the undrained constraint can be established by integrating the constitutive relation (3.14). For $d\varepsilon_v \to 0$ there is

$$dq = \left(3G - \frac{9G^2}{K\eta_f^2 + 3G}\right) d\varepsilon_q \qquad (3.19)$$

so that

$$q = q_0 + \left(3G - \frac{9G^2}{K\eta_f^2 + 3G}\right)(\varepsilon_q - \varepsilon_q^0) \qquad (3.20)$$

where q_0, ε_q^0 are the values at the onset of plastic deformation, Figure 3.3b.

The material behaviour as depicted in Figure 3.3 is not commonly observed when conducting standard 'triaxial' tests on soils. In fact, the mechanical characteristics under undrained conditions are quite complex and strongly depend on the initial porosity of the material. The next section provides a review of basic trends in the behaviour of cohesionless granular soils. This review in itself supplies quite convincing arguments for employing more refined approaches than a simple perfectly plastic idealization.

3.1.3 Review of typical mechanical characteristics of granular materials

Let us briefly review the basic trends in the 'triaxial' response of cohesionless granular materials, such as sand. In general, the mechanical characteristics, under both drained and undrained conditions, are strongly influenced by the initial porosity/void ratio. The quantitative trends are depicted in a series of figures 3.4–3.7.

Consider first the behaviour under drained conditions. Figure 3.4 shows the characteristics which correspond to axial compression at some initial confining pressure. In general, in loose sand, the mechanical response is stable (in Drucker's sense). An increase in the deviatoric stress is associated with progressive *compaction* (decrease in volume); as the conditions at failure are approached the volume change tends to zero,

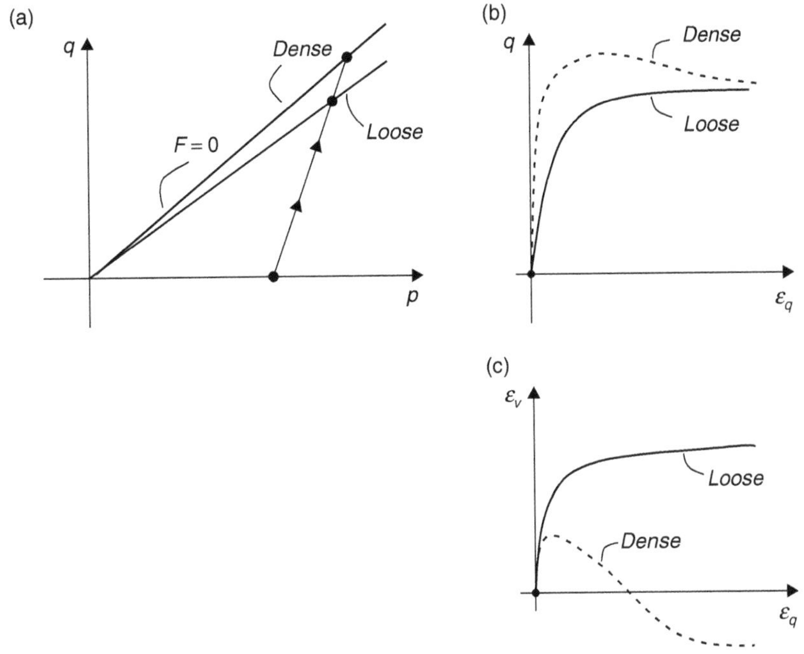

Figure 3.4 Typical response of a granular material subjected to drained axial compression

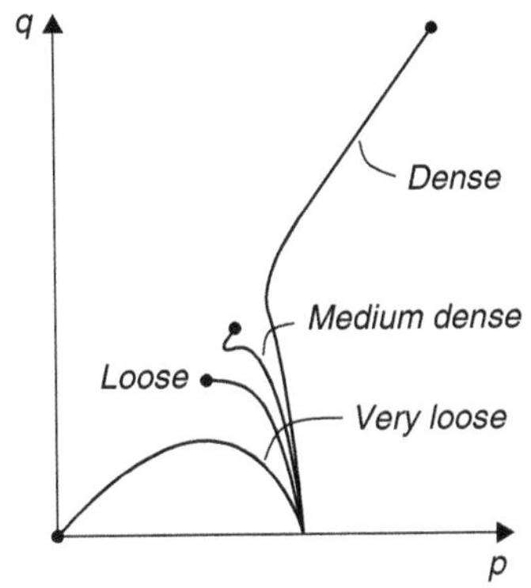

Figure 3.5 Typical effective stress trajectories under undrained constraint

Figure 3.4c. The same sand, tested under a different initial void ratio, will display different characteristics. In general, dense sand has a higher compressive strength as well as a higher initial stiffness, Figure 3.4b. The deviatoric characteristic is stable prior to reaching the peak. The actual failure mechanism is typically associated with formation of a shear band which triggers a strain-softening response (descending branch) at the macroscale. At the initial stage of the deformation process, the material undergoes compaction, Figure 3.4c. At some instant, a transition to *dilation* (i.e. increase in volume) occurs. Again, as the conditions at failure are approached, the volume change tends to zero.

The behaviour under drained conditions has significant implications on the behaviour under undrained constraint. In general, if the material displays a tendency to compaction (loose sand), this leads to a progressive build up of the excess of pore pressure. Conversely, in a dilating material (dense sand) a generation of negative excess of pore pressure takes place.

Figure 3.5 shows the effective stress trajectories for a broad range of initial void ratios. In *dense* sand, the plastic dilation triggers a build up of negative excess of pore pressure. Consequently, the effective stress trajectory migrates away from the stress space origin and gradually approaches the failure envelope. On the other hand, in *very loose* sand, a significant generation of positive pore pressure takes place. At a certain stage, a loss of stability occurs associated with decrease in the deviatoric stress intensity under continuing deformation. At the end of the test the effective pressure reduces to zero and the material *liquefies*. Apparently, the stress trajectory corresponding to *loose/medium dense* sand is an intermediate between the two extreme cases.

The notion of *liquefaction* deserves perhaps an additional comment. In general, the liquefaction should be understood as a state within the sample at which the mean effective pressure reduces to zero, i.e. $p_w = \hat{p}; p = 0$. At this point, the contact between grains is lost and the material behaves as a *viscous fluid*. In the field situations, the liquefaction is typically triggered by seismic excitation and may occur in loose as well as medium dense sand. It often has a devastating effect, leading to collapse of engineering structures and a loss of life. In order to fully appreciate the conditions for liquefaction, let us briefly review the basic trends in the response of granular media under cyclic loading.

Figure 3.6 shows a typical volume change characteristic corresponding to drained conditions. When sample that is in a relatively loose state of compaction is subjected to a series of stress cycles with low/moderate amplitudes, a progressive decrease in volume (compaction) occurs. As the cycles continue, the rate of compaction reduces leading eventually to a stationary state. As mentioned earlier, the irreversible reduction in volume results in a build up of positive pore pressure under undrained conditions. The implications of this are evident in Figure 3.7, which shows typical effective stress trajectories for loose/medium dense and dense sand, respectively. In the former case, Figure 3.7a, a progressive generation of pore pressure takes place, which triggers a migration of the effective stress trajectory towards the stress space origin. After a number of cycles the effective pressure reduces to zero, which signifies *liquefaction* of the material. In dense sand, Figure 3.7b, an initial build up of pore pressure is also observed. After several cycles, however, the dilatancy effects become prominent leading to generation of negative excess of pore pressure. As a result, the stress trajectory traces a closed loop in the effective stress space and the liquefaction does not occur. This phenomenon is commonly referred to as *cyclic mobility*.

The overview, as presented above, highlights the basic trends in the mechanical response of saturated granular media. A more comprehensive assessment is provided in Chapter 8, which deals with the actual experimental data. It is evident that the mechanical characteristics are quite complex and diversified; so that an elastic-perfectly plastic idealization is not fully adequate and more refined approaches are, in general, required.

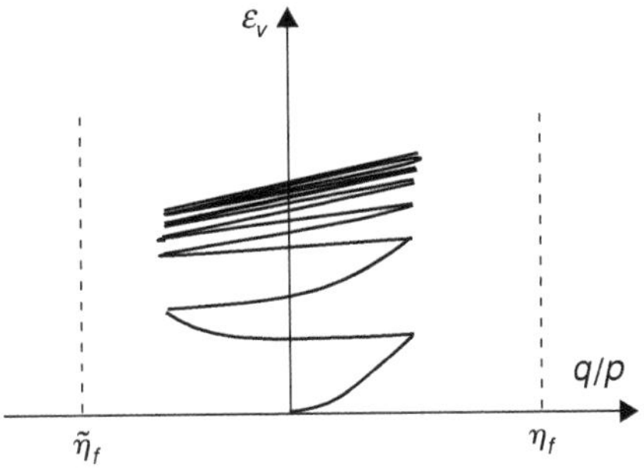

Figure 3.6 Volume change characteristic in a drained cyclic test (loose/medium dense sand)

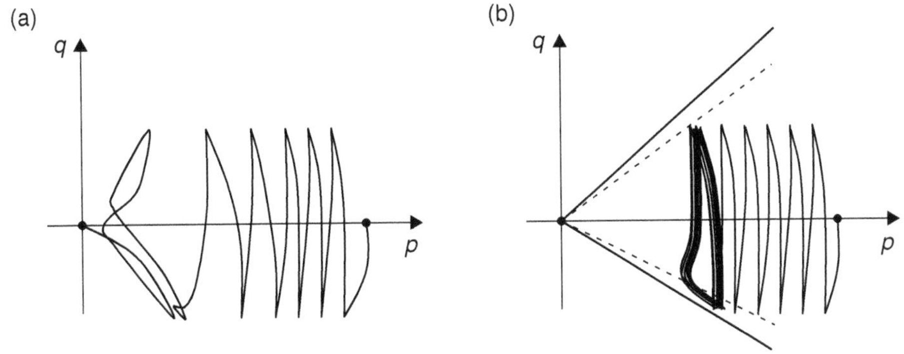

Figure 3.7 Typical response of (a) loose/medium dense and (b) dense sand in an undrained stress-controlled cyclic test

3.2 VOLUMETRIC HARDENING; CRITICAL STATE MODEL

Since most of the experimental evidence comes from 'triaxial' tests, it is convenient to develop first a simplified $\{p, q\}$ formulation and later generalize it for an arbitrary stress state. The framework outlined in this section employs isotropic-hardening and attributes the hardening effects to accumulated irreversible volume change. This formulation is commonly referred to as *Critical State* approach [24,25] and it is well established by now in the geotechnical research community.

3.2.1 Formulation in the 'triaxial' {p,q} space

Within the framework of elastic–perfectly plastic material, the conditions at failure are said to be path-independent and the failure criterion takes a general form $F = F(p, q) = 0$. In contrast to this approach, the concept of volumetric hardening employs the notion of a loading surface, whose evolution depends on a hardening parameter; the latter identified with irreversible void ratio e^p. Thus,

$$f = f(p, q, e^p) = 0 \tag{3.21}$$

where

$$de^p = -(1 + e_0)d\varepsilon_v^p \tag{3.22}$$

and e_0 is the initial void ratio.

The approach typically involves an associated flow rule, i.e.

$$d\varepsilon_v^p = d\lambda \frac{\partial f}{\partial p}; \quad d\varepsilon_q^p = d\lambda \frac{\partial f}{\partial q} \tag{3.23}$$

The constitutive relation is derived by imposing the consistency condition

$$df = \frac{\partial f}{\partial p} dp + \frac{\partial f}{\partial q} dq + \frac{\partial f}{\partial e^p} de^p = 0 \tag{3.24}$$

where

$$dp = K(d\varepsilon_v - d\varepsilon_v^p); \quad dq = 3G(d\varepsilon_q - d\varepsilon_q^p) \tag{3.25}$$

Substituting now eqs.(3.25), together with (3.22) and (3.23), in the consistency condition (3.24) yields

$$d\lambda = \frac{\dfrac{\partial f}{\partial p} K\, d\varepsilon_v + \dfrac{\partial f}{\partial q} 3G\, d\varepsilon_q}{H_e + H_p} \tag{3.26}$$

where

$$H_e = K\left(\frac{\partial f}{\partial p}\right)^2 + 3G\left(\frac{\partial f}{\partial q}\right)^2; \quad H_p = (1 + e_0)\frac{\partial f}{\partial e^p}\frac{\partial f}{\partial p} \tag{3.27}$$

and H_p is referred to as the *plastic hardening modulus*.

For any *strain-controlled* program, the general form of the constitutive relation can now be obtained by substituting the expression for the plastic multiplier (3.26) in the elasticity equations (3.25). This leads to

$$\begin{Bmatrix} dp \\ dq \end{Bmatrix} = \begin{bmatrix} \left(K - \dfrac{\dfrac{\partial f}{\partial p} K^2 \dfrac{\partial f}{\partial p}}{H_e + H_p} \right) & \left(- \dfrac{\dfrac{\partial f}{\partial q} 3GK \dfrac{\partial f}{\partial p}}{H_e + H_p} \right) \\ \left(- \dfrac{\dfrac{\partial f}{\partial p} 3GK \dfrac{\partial f}{\partial q}}{H_e + H_p} \right) & \left(3G - \dfrac{\dfrac{\partial f}{\partial q} 9G^2 \dfrac{\partial f}{\partial q}}{H_e + H_p} \right) \end{bmatrix} \begin{Bmatrix} d\varepsilon_v \\ d\varepsilon_q \end{Bmatrix} \tag{3.28}$$

Conversely, in a typical *stress-controlled* program, the consistency condition (3.24) together with (3.22) results in

$$d\lambda = \frac{1}{H_p}\left(\frac{\partial f}{\partial p} dp + \frac{\partial f}{\partial q} dq \right); \quad H_p = (1 + e_0) \frac{\partial f}{\partial e^p} \frac{\partial f}{\partial p} \tag{3.29}$$

Thus, invoking the additivity of elastic and plastic strain increments, the inverse relation to that in (3.28) may be expressed in the form

$$\begin{Bmatrix} d\varepsilon_v \\ d\varepsilon_q \end{Bmatrix} = \begin{bmatrix} \left(\dfrac{1}{K} + \dfrac{1}{H_p} \left(\dfrac{\partial f}{\partial p} \right)^2 \right) & \left(\dfrac{1}{H_p} \dfrac{\partial f}{\partial q} \dfrac{\partial f}{\partial p} \right) \\ \left(\dfrac{1}{H_p} \dfrac{\partial f}{\partial q} \dfrac{\partial f}{\partial p} \right) & \left(\dfrac{1}{3G} + \dfrac{1}{H_p} \left(\dfrac{\partial f}{\partial q} \right)^2 \right) \end{bmatrix} \begin{Bmatrix} dp \\ dq \end{Bmatrix} \tag{3.30}$$

It should be noted that both the stiffness and compliance operators, as defined in eqs.(3.28) and (3.30), are positive definite. Clearly the stiffness operator becomes singular for $H_p = 0$, i.e. when the conditions at failure are reached and the material becomes perfectly plastic.

The constitutive relation (3.28) or (3.30) provides a general formulation of the problem. The implementation of this framework requires the specification of the loading function $f = 0$, eq.(3.21), and its evolution in the course of plastic deformation. In order to accomplish this, consider first the definition of the plastic hardening modulus, viz. eq.(3.29). It is evident here that

$$H_p = 0 \quad \Leftrightarrow \quad \frac{\partial f}{\partial p} = 0 \quad \Rightarrow \quad d\varepsilon_v^p = 0 \tag{3.31}$$

i.e., failure is combined with no irreversible volume change. The conditions at failure are typically defined by Mohr-Coulomb criterion $F = 0$, eq.(3.8). Taking, for simplicity, a cohesionless material, one can write

$$F = 0 \quad \Rightarrow \quad q = \eta_f p; \quad \eta_f = 6 \sin\phi / (3 - \sin\phi) \quad (\textit{for } q > 0) \tag{3.32}$$

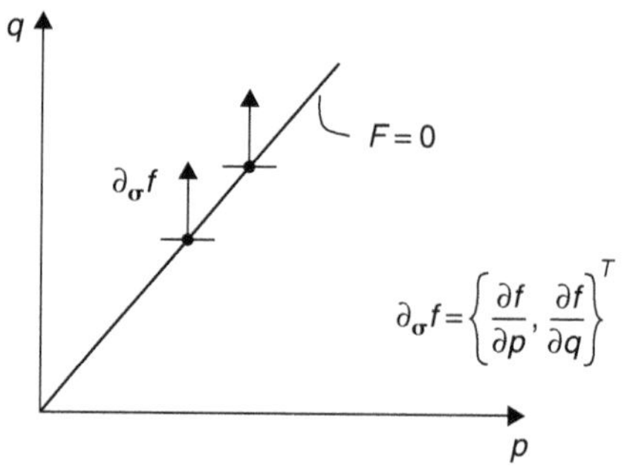

Figure 3.8 Admissibility condition for the loading surface

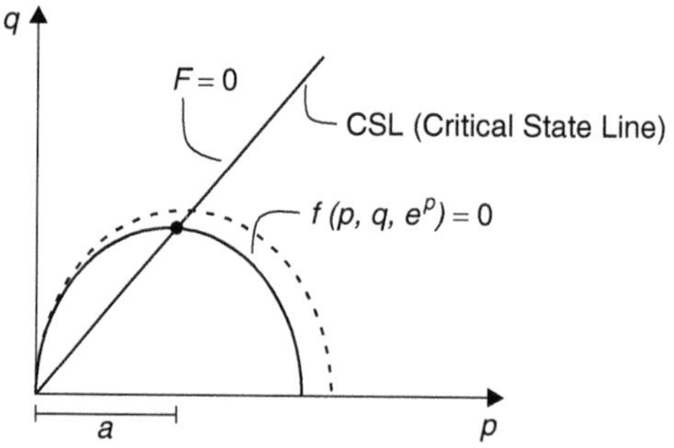

Figure 3.9 Volumetric hardening; an elliptical yield surface

Thus, any surface, which is convex and satisfies $q=\eta_f p \Leftrightarrow \partial f/\partial p=0$ may be considered an admissible loading surface, Figure 3.8. As an example, an elliptical surface may be employed

$$f(p,q,e^p) = (p-a)^2 + \left(\frac{q}{\eta_f}\right)^2 - a^2 = 0; \quad a = a(e^p) \tag{3.33}$$

as shown in Figure 3.9. Note that since in extension domain $\eta_f := \tilde{\eta}_f$, the loading surface is not symmetric about the p-axis.

The formulation incorporating (3.33) is known as modified *Cam Clay* model. Here, the failure envelope $F=0$ is commonly referred to as the *Critical State line* (CSL), and it is defined by the conditions $H_p=0 \wedge d\varepsilon_v^p=0$. Furthermore, according to eq.(3.33)

$$\frac{\partial f}{\partial p} = 2(p-a); \quad \frac{\partial f}{\partial e^p} = -2p\,\frac{\partial a}{\partial e^p} \tag{3.34}$$

so that $p \to a \Rightarrow H_p \to 0$ and $a=a(e^p)$ is a material function which needs to be determined experimentally.

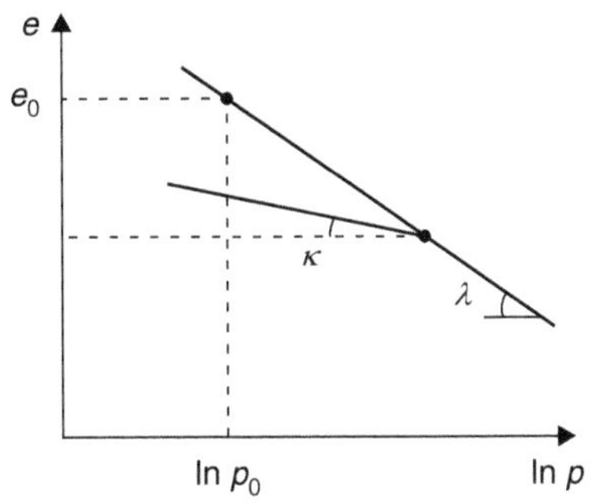

Figure 3.10 Typical results of a hydrostatic compression test

The function $a(e^p)$ is commonly identified by conducting a hydrostatic compression test. Typical results, for a normally consolidated clay, are shown schematically in Figure 3.10. The characteristics are plotted in the affine $\{e, \ln p\}$ plane, where e is the void ratio. During the virgin loading the relation remains linear; the slope of the consolidation line is equal to λ which is commonly referred to as *compression index*. Unloading path produces a similar linear response with a slope of κ, known as the *swelling index*[2]. Let the initial void ratio, say at $p = p_0$, be $e = e_0$. Given the linear relation in Figure 3.10, the evolution of e, as measured from the initial state e_0, can be described as

$$e = e_0 - \lambda \ln \frac{p}{p_0}; \quad e^e = e_0 - \kappa \ln \frac{p}{p_0} \tag{3.35}$$

where the second equation defines the elastic, reversible changes. Assuming now the additivity of elastic and plastic volume changes, we have

$$e^p = e - e^e = (\kappa - \lambda) \ln \frac{p}{p_0} \quad \Rightarrow \quad p = p_0 \exp\left(\frac{-e^p}{\lambda - \kappa}\right) \tag{3.36}$$

Now, since for hydrostatic compression there is $a = 2p$, Figure 3.9, the function $a = a(e^p)$ can be defined in a way consistent with representation (3.36), i.e.

$$a = a_0 \exp\left(\frac{-e^p}{\lambda - \kappa}\right) \quad \Rightarrow \quad H_p = (1 + e_0) \frac{4ap(p-a)}{\lambda - \kappa} \tag{3.37}$$

which, in turn, also defines the plastic hardening modulus H_p, eq.(3.29).

---

[2] Both λ and κ have been used before. The present notation, however, is standard in Critical State theory, so the same symbols are employed in a different context

Finally, it is noted that differentiating the second equation in (3.35), i.e. the one that governs the elastic volume change, yields

$$dp = \left(\frac{1+e_0}{\kappa} p\right) d\varepsilon_v^e = K\, d\varepsilon_v^e \quad \Rightarrow \quad K = \frac{1+e_0}{\kappa} p \qquad (3.38)$$

Thus, the linear response in $e - \ln p$ plane is, in fact, equivalent to a non-linear elastic representation, in which the bulk modulus K remains proportional to the hydrostatic pressure.

The formulation, as presented above, is now complete. The framework incorporates four basic material constants; G, κ which define the elastic properties, the angle of internal friction ϕ that specifies the conditions at failure, and the compression index λ which defines the hardening characteristics. All constants can be identified from standard 'triaxial' tests.

3.2.2 Comments on the performance

Let us examine now the basic trends in the mechanical response as stipulated by the Critical State model. For this purpose choose some arbitrary material parameters, that may be considered as typical for a class of cohesionless soils, e.g.

$$G = 30\,\text{MPa}, \quad \phi = 30^0, \quad \lambda = 0.13, \quad \kappa = 0.02, \quad e_0 = 0.9$$

Consider first the response under *drained* conditions, as depicted in Figure 3.11. Here, the sample is assumed to be consolidated under hydrostatic conditions and subsequently failed under axial compression. Different initial conditions are simulated, involving a normally consolidated material (confinement of $p_0 = 500\,\text{kPa}$), as well as an *overconsolidated* one. In the latter case, the initial confinement of 500 kPa is reduced to 300 kPa and 100 kPa, respectively, prior to the increase in the vertical stress. The mechanical characteristics are shown in Figure 3.11b. In normally consolidated material, the plastic deformation develops from the onset of loading and the increase in deviatoric stress is associated with progressive compaction, Figure 3.11c. In lightly overconsolidated material ($p_0 = 300\,\text{kPa}$, i.e. OCR = 1.67), the behaviour is elastic prior to reaching the loading surface. In the plastic regime, the increase in the vertical stress is, once again, coupled with a progressive decrease in volume. In case of a higher overconsolidation ratio ($p_0 = 100\,\text{kPa}$, i.e. OCR = 5), the stress trajectory reaches the loading surface in the region where $H_p < 0$, eq.(3.37), implying an unstable strain-softening response associated with continuing dilation, Figure 3.11c.

It needs to be emphasized again that the strain softening behaviour typically involves an *inhomogeneous* deformation mode triggered by formation of a shear band. Thus, the descending branch, as depicted in Figure 3.11b, does not represent the *material* response but that of a structural system. In fact, the rate of strain softening will now be significantly affected by the geometry of the sample. Therefore, the range of applicability of the Critical State model should be perceived as being

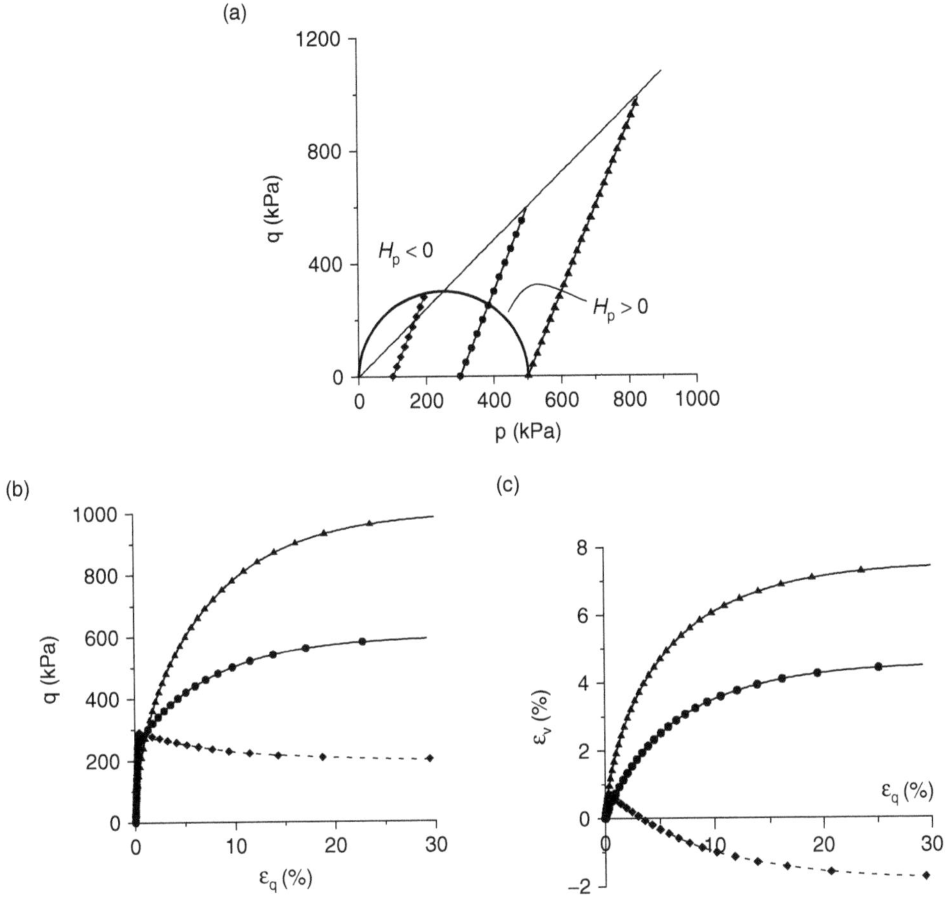

Figure 3.11 Critical State model; numerical simulations of a series of drained axial compression tests

restricted to the domain in which $H_p > 0$, i.e. the material displays strain-hardening characteristics.

Consider now the response under *undrained* conditions. The constraint of no volume change results in

$$d\varepsilon_v = 0 \quad \Rightarrow \quad d\varepsilon_v^e = -d\varepsilon_v^p = \frac{dp}{K} \tag{3.39}$$

where, according to eq.(3.29)

$$d\varepsilon_v^p = \frac{1}{H_p}\left(\frac{\partial f}{\partial p}dp + \frac{\partial f}{\partial q}dq\right)\frac{\partial f}{\partial p} \tag{3.40}$$

For an elliptical surface, eqs.(3.33) and (3.37), there is

$$d\varepsilon_v^p = \frac{(p-a)dp + q\eta_f^{-2}dq}{(1+e_0)\dfrac{ap}{\lambda-\kappa}} = -\frac{dp}{K} \tag{3.41}$$

so that the *effective stress trajectory* is defined by a differential form

$$\frac{dq}{dp} = -\frac{p^2\eta_f^2(\beta+1) + q^2(\beta-1)}{2qp}; \quad \beta = \frac{\kappa}{\lambda-\kappa} \tag{3.42}$$

Integrating (3.42), subject to the initial condition $q(p_0)=0$, gives

$$q = \eta_f p^{\frac{1}{2}(1-\beta)} \sqrt{p_0^{(1+\beta)} - p^{(1+\beta)}} \tag{3.43}$$

Figure 3.12 shows the results for a normally consolidated sample subjected to initial confining pressure of $p_0 = 500$ kPa. Since the material undergoes plastic compaction, the effective stress path (E) migrates towards the stress space origin, thereby resulting in generation of positive pore pressures. Overall, the trends as depicted in Figures 3.11 and 3.12 indicate that the range of applicability of the Critical State formulation extends primarily to normally consolidated and/or lightly overconsolidated clays as well as sand, in a relatively loose state of compaction.

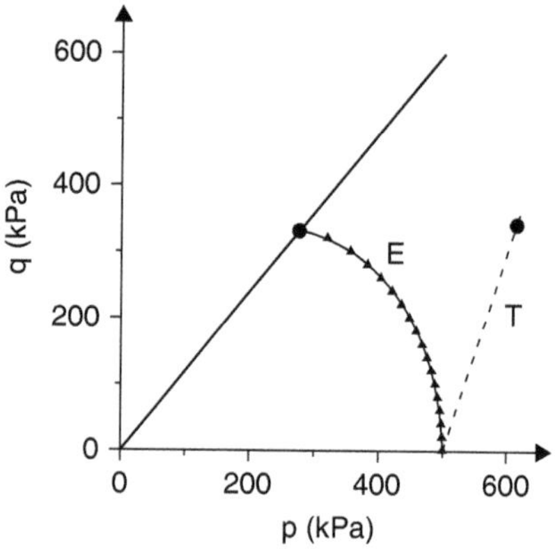

Figure 3.12 Critical State model; stress trajectories under undrained conditions

3.2.3 Generalization and specification of the constitutive matrix

The generalization of the functional form of the loading surface is based on representation (2.25) combined with expressions (3.2) and (3.3); the latter defining the stress invariants in terms of parameters p and q. Thus,

$$p = \sigma_m; \quad |q| = \sqrt{3}\bar{\sigma}/g(\theta) \tag{3.44}$$

where $g(\theta)$ defines the shape of the loading surface in the deviatoric plane and can be taken in the form consistent with either (2.23) or (2.41)–(2.43).

Substituting the above relation in the expression (3.33) results in

$$f = (\sigma_m - a)^2 + \left(\frac{\sqrt{3}\bar{\sigma}}{\eta_f g(\theta)}\right)^2 - a^2 = 0 \tag{3.45}$$

Apparently, $g(\theta) = const.$ will give an ellipsoid; choosing $g(\theta)$ in one of the forms discussed in Section 2 will result in octahedral sections that are smooth approximations to Mohr-Coulomb hexagon, Figure 3.13.

The general formulation of the problem follows now the procedure as outlined in Section 2.5. The consistency condition reads

$$f(\boldsymbol{\sigma}, e^p) = 0; \quad df = \left(\frac{\partial f}{\partial \boldsymbol{\sigma}}\right)^T d\boldsymbol{\sigma} + \frac{\partial f}{\partial e^p} de^p = 0 \tag{3.46}$$

Invoking the additivity postulate, the response in the elastic range is defined by

$$d\boldsymbol{\varepsilon} = d\boldsymbol{\varepsilon}^e + d\boldsymbol{\varepsilon}^p \quad \Rightarrow \quad d\boldsymbol{\sigma} = [D^e](d\boldsymbol{\varepsilon} - d\boldsymbol{\varepsilon}^p) \tag{3.47}$$

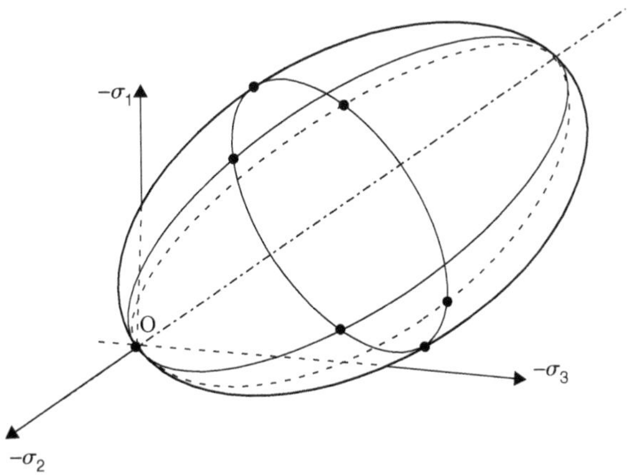

Figure 3.13 Volumetric hardening; loading surface in the principal stress space

while the plastic strain increments satisfy

$$d\boldsymbol{\varepsilon}^p = d\lambda \frac{\partial f}{\partial \boldsymbol{\sigma}}; \quad de^p = -(1+e_0)\boldsymbol{\delta}^T d\boldsymbol{\varepsilon}^p \tag{3.48}$$

where, according to notation employed in Section 2.5.1, $\boldsymbol{\delta} = \{1,1,1,0,0,0\}^T$.

Substituting the above equations in the consistency condition (3.46) yields

$$\left(\frac{\partial f}{\partial \boldsymbol{\sigma}}\right)^T [D^e] d\boldsymbol{\varepsilon} - \left(\frac{\partial f}{\partial \boldsymbol{\sigma}}\right)^T [D^e]\left(\frac{\partial f}{\partial \boldsymbol{\sigma}}\right) d\lambda - \frac{\partial f}{\partial e^p}(1+e_0)\boldsymbol{\delta}^T \frac{\partial f}{\partial \boldsymbol{\sigma}} d\lambda = 0 \tag{3.49}$$

Solving now for the plastic multiplier

$$d\lambda = \frac{1}{H}\left(\frac{\partial f}{\partial \boldsymbol{\sigma}}\right)^T [D^e] d\boldsymbol{\varepsilon}; \quad H = H_e + H_p$$
$$H_e = \left(\frac{\partial f}{\partial \boldsymbol{\sigma}}\right)^T [D^e] \frac{\partial f}{\partial \boldsymbol{\sigma}}; \quad H_p = \frac{\partial f}{\partial e^p}(1+e_0)\boldsymbol{\delta}^T \frac{\partial f}{\partial \boldsymbol{\sigma}} \tag{3.50}$$

Thus, the constitutive relation takes the form

$$d\boldsymbol{\sigma} = [D^{ep}] d\boldsymbol{\varepsilon}, \quad [D^{ep}] = [D^e] - \frac{1}{H}\left([D^e]\frac{\partial f}{\partial \boldsymbol{\sigma}}\left(\frac{\partial f}{\partial \boldsymbol{\sigma}}\right)^T [D^e]\right) \tag{3.51}$$

The above expression, together with representation (3.45), defines the general form of the constitutive matrix. It is noted that the relation (3.51) is similar to (2.72), as derived in the previous chapter; this time, however, $[D^{ep}]$ is a positive-definite matrix, as long as $H_p > 0$.

3.3 DEVIATORIC HARDENING MODEL

In the Critical State framework, as discussed in the previous section, the hardening effects are attributed to irreversible volume change. In this section, a different approach is presented which incorporates the notion of a deviatoric hardening (cf. Ref.[26]). Thus, the hardening effects are ascribed to development of plastic distortions. Again, a simplified $\{p,q\}$ formulation is outlined first followed by generalization suitable for applications involving 3D problems.

3.3.1 Formulation in the 'triaxial' {p,q} space

Within the framework of the deviatoric hardening, the loading surface assumes the form

$$f(p, q, \varepsilon_q^p) = 0 \tag{3.52}$$

where ε^p_q is the accumulated deviatoric plastic strain. The formulation incorporates a non-associated flow rule, i.e.

$$d\varepsilon^p_v = d\lambda\frac{\partial\psi}{\partial p};\quad d\varepsilon^p_q = d\lambda\frac{\partial\psi}{\partial q};\quad \wedge\ \psi = \psi(p,q) = const. \tag{3.53}$$

where ψ is the plastic potential function.

The constitutive relation is derived by imposing the consistency condition

$$df = \frac{\partial f}{\partial p}dp + \frac{\partial f}{\partial q}dq + \frac{\partial f}{\partial \varepsilon^p_q}d\varepsilon^p_q = 0 \tag{3.54}$$

where

$$dp = K(d\varepsilon_v - d\varepsilon^p_v);\quad dq = 3G(d\varepsilon_q - d\varepsilon^p_q) \tag{3.55}$$

Substituting eqs.(3.55) and (3.53) in (3.54) yields

$$d\lambda = \frac{\dfrac{\partial f}{\partial p}K\,d\varepsilon_v + \dfrac{\partial f}{\partial q}3G\,d\varepsilon_q}{H_e + H_p} \tag{3.56}$$

where

$$H_e = \frac{\partial f}{\partial p}K\frac{\partial\psi}{\partial p} + \frac{\partial f}{\partial q}3G\frac{\partial\psi}{\partial q};\quad H_p = -\frac{\partial f}{\partial\varepsilon^p_q}\frac{\partial\psi}{\partial q} \tag{3.57}$$

Thus, for any *strain-controlled* program, the general form of the constitutive relation becomes

$$\begin{Bmatrix} dp \\ dq \end{Bmatrix} = \begin{bmatrix} \left(K - \dfrac{\dfrac{\partial f}{\partial p}K^2\dfrac{\partial\psi}{\partial p}}{H_e + H_p}\right) & \left(-\dfrac{\dfrac{\partial f}{\partial q}3GK\dfrac{\partial\psi}{\partial p}}{H_e + H_p}\right) \\ \left(-\dfrac{\dfrac{\partial f}{\partial p}3GK\dfrac{\partial\psi}{\partial q}}{H_e + H_p}\right) & \left(3G - \dfrac{\dfrac{\partial f}{\partial q}9G^2\dfrac{\partial\psi}{\partial q}}{H_e + H_p}\right) \end{bmatrix} \begin{Bmatrix} d\varepsilon_v \\ d\varepsilon_q \end{Bmatrix} \tag{3.58}$$

For a *stress-controlled* program, the plastic multiplier can be defined directly from (3.54) by employing the non-associated flow rule (3.53)

$$d\lambda = \frac{1}{H_p}\left(\frac{\partial f}{\partial p}dp + \frac{\partial f}{\partial q}dq\right);\quad H_p = -\frac{\partial f}{\partial\varepsilon^p_q}\frac{\partial\psi}{\partial q} \tag{3.59}$$

Given now eq.(3.59), the inverse relation to that in (3.58) can be derived by assuming the additivity of elastic and plastic strain increments. Thus,

$$\begin{Bmatrix} d\varepsilon_v \\ d\varepsilon_q \end{Bmatrix} = \begin{bmatrix} \left(\frac{1}{K} + \frac{1}{H_p}\frac{\partial f}{\partial p}\frac{\partial \psi}{\partial p}\right) & \left(\frac{1}{H_p}\frac{\partial f}{\partial q}\frac{\partial \psi}{\partial p}\right) \\ \left(\frac{1}{H_p}\frac{\partial f}{\partial p}\frac{\partial \psi}{\partial q}\right) & \left(\frac{1}{3G} + \frac{1}{H_p}\frac{\partial f}{\partial q}\frac{\partial \psi}{\partial q}\right) \end{bmatrix} \begin{Bmatrix} dp \\ dq \end{Bmatrix} \tag{3.60}$$

It is evident that both the stiffness and compliance operators, as defined above, are non-symmetric, in view of the non-associated flow rule.

In order to complete the general formulation, as stipulated by (3.58) or (3.60), the functional form of the loading and plastic potential functions needs to be specified. Consider, for simplicity, a cohesionless material. In compression domain, i.e. for $q > 0$, the loading surface is typically chosen in a simple linear form

$$f = q - \eta p = 0; \quad \eta = \eta(\varepsilon_q^p) \tag{3.61}$$

The hardening function $\eta = \eta(\varepsilon_q^p)$ is defined within the interval $\eta \in [0, \eta_f)$ and is selected in a hyperbolic form

$$\eta = \eta_f \frac{\varepsilon_q^p}{A + \varepsilon_q^p} \tag{3.62}$$

where A is a material constant and η_f defines the conditions at failure. The latter are assumed to be consistent with Mohr-Coulomb criterion, so that

$$\varepsilon_q^p \to \infty \quad \Rightarrow \quad \eta \to \eta_f = \frac{6 \sin \phi}{3 - \sin \phi} \tag{3.63}$$

where ϕ is the angle of internal friction.

The plastic potential function is chosen in such a way as to ensure a smooth transition from compaction to dilatancy prior to failure. This can be achieved by selecting the plastic potential in the form similar to that employed in the original Cam-Clay formulation [24], i.e.

$$\psi = q + \eta_c\, p \ln\left(\frac{p}{\bar{p}}\right) = 0 \tag{3.64}$$

The above expression represents a parametric equation in which $\eta_c = const.$ and $\bar{p}$ is evaluated from the condition of $\psi(p, q) = 0$. It is noted that according to eq.(3.64)

$$\frac{\partial \psi}{\partial p} = \eta_c\left(\ln\frac{p}{\bar{p}} + 1\right) = \eta_c - \eta \tag{3.65}$$

so that $\eta = \eta_c \Rightarrow \partial\psi/\partial p = 0 \Rightarrow d\varepsilon_v^p = 0$. The locus $\eta = q/p = \eta_c$ is referred to as *zero dilatancy line*. Clearly, for $\eta < \eta_c$ there is $d\varepsilon_v^p > 0$, which implies *compaction*, whereas $\eta_f > \eta > \eta_c \Rightarrow d\varepsilon_v^p < 0$, which corresponds to *dilation*.

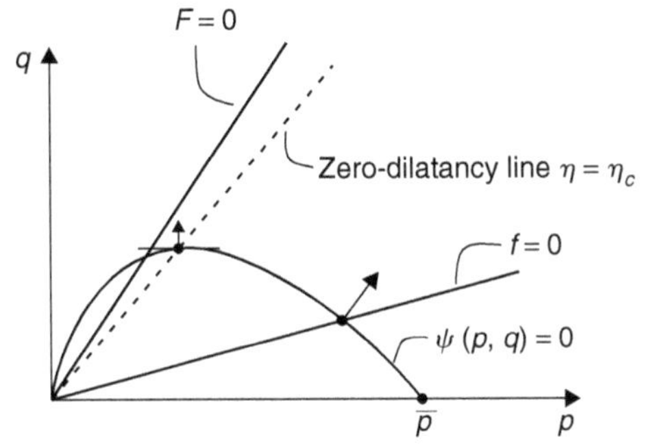

Figure 3.14 Schematic representation of deviatoric hardening framework

Given the definitions above, the plastic hardening modulus H_p, eq.(3.57), can be expressed as

$$H_p = -\frac{\partial f}{\partial \varepsilon^p_q}\frac{\partial \psi}{\partial q} = \eta' p; \quad \eta' = \frac{d\eta}{d\varepsilon^p_q} = \frac{(\eta_f - \eta)^2}{A\eta_f} \tag{3.66}$$

Thus $H_p \to 0$ requires $\eta \to \eta_f$, which again is consistent with the Mohr-Coulomb criterion.

Figure 3.14 shows a schematic representation of the deviatoric hardening framework in the compression domain of the $\{p,q\}$-space. Note again that in extension $\eta_f := \tilde{\eta}_f$; $\eta_c := \tilde{\eta}_c$, so that both the loading and plastic potential surfaces are asymmetric. The formulation incorporates five basic parameters: elastic constants G, K; the angle of internal friction ϕ which, in turn, defines the slope of the failure envelope η_f; the parameter η_c which specifies the slope of zero dilatancy line and the constant A which appears in the hardening function (3.62). Once again, all parameters can be identified from standard triaxial tests.

3.3.2 Comments on the performance

Let us examine now the basic trends in the mechanical response as predicted by the deviatoric hardening model. For this purpose choose some material parameters that may be considered as representative for granular soils, like sand. Let us focus first on material which is in a dense state of compaction and assume

$$G = 45\,\text{MPa}; \quad K = 100\,\text{MPa}; \quad \eta_f = 1.4; \quad \eta_c = 1.2; \quad A = 0.0005$$

Figure 3.15 presents the results of numerical simulations for a *drained* axial compression test conducted at the initial confining pressure of 500 kPa. It is evident that as the axial stress increases, a progressive transition from compaction to dilatancy takes place. When the conditions at failure are approached the rate of dilation becomes constant, Figure 3.15c. The latter aspect is not entirely consistent with the experimental

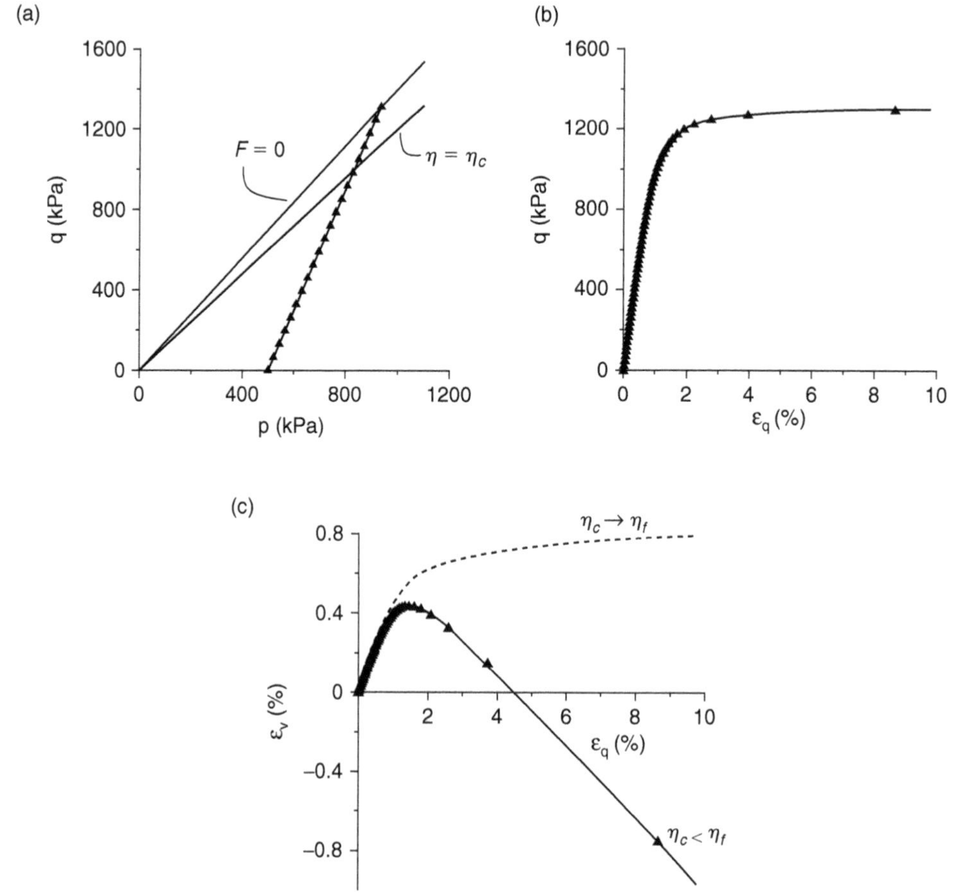

Figure 3.15 Deviatoric hardening; numerical simulations of a drained axial compression test

evidence, which indicates that the volume change should progressively approach zero, Figure 3.4c. In general, the amount of dilation is governed primarily by the ratio η_f/η_c. When $\eta_c \rightarrow \eta_f$, the material undergoes a continuing compaction, Figure 3.15c.

It is also worth noting that, within the present framework, all trajectories involving $\eta = const.$, including isotropic compression, are neutral, i.e. produce no plastic deformations. The latter contradicts, in general, the existing experimental data.

Figure 3.16 shows the results of numerical simulations for *undrained* axial compression at the same initial confining pressure of 500 kPa. The characteristics correspond to different initial degrees of compaction, ranging from dense to very loose configurations. In *dense* sample, Figure 3.16a, a generation of negative excess of pore pressure takes place and the effective stress trajectory approaches the failure envelope. Note that in this case there is $\eta_c = 0.86\eta_f$, so that a significant plastic dilation occurs as $\eta \rightarrow \eta_f$.

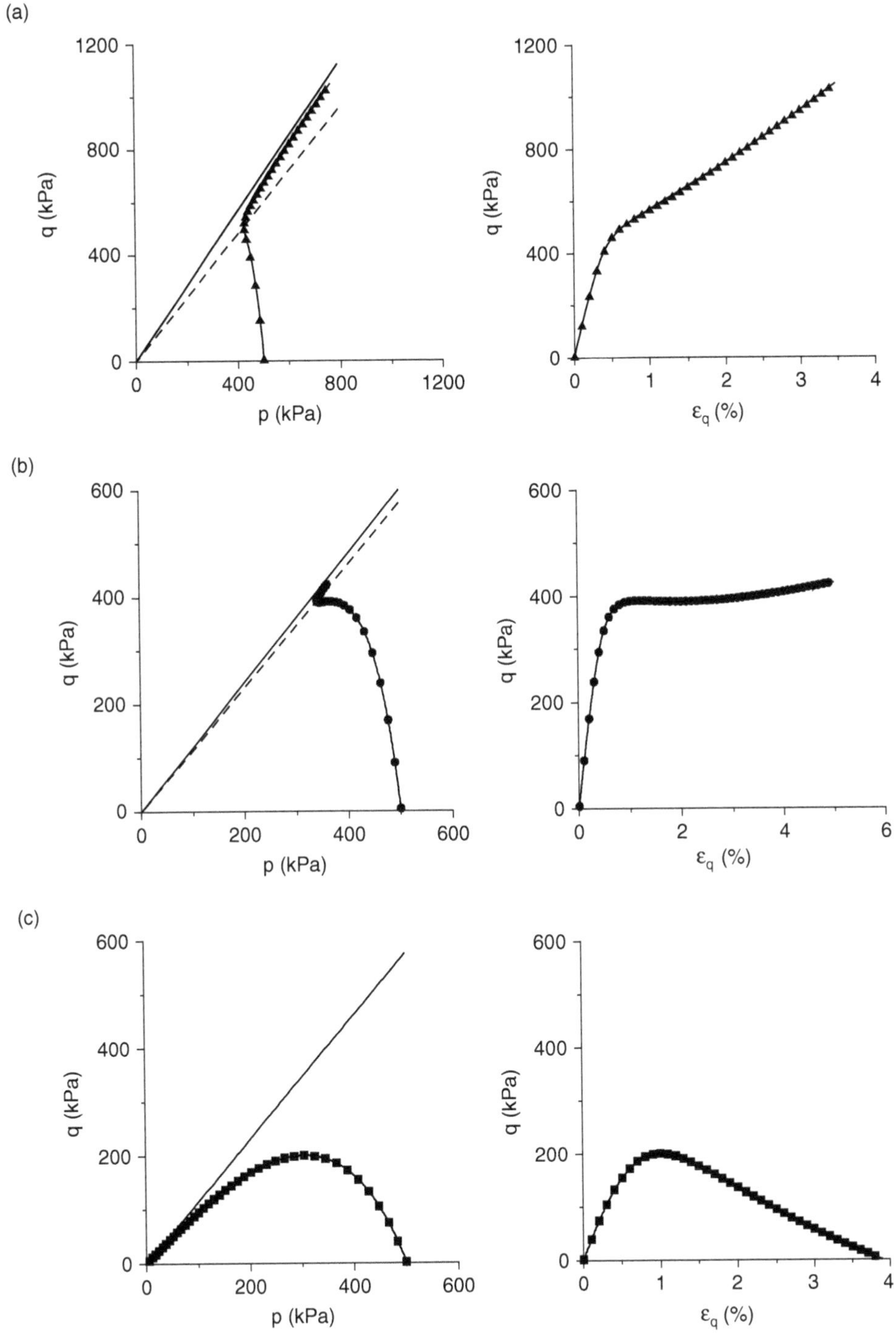

Figure 3.16 Deviatoric hardening; numerical simulations of a series of undrained compression tests

The results presented in Figure 3.16b correspond to

$$G = 35\,\text{MPa}; \quad K = 60\,\text{MPa}; \quad \eta_f = 1.2; \quad \eta_c = 1.15; \quad A = 0.001$$

which may be perceived as a set of parameters representative of *loose/medium dense* sand. In this case η_c is only marginally smaller than η_f. Consequently, the predominant mode is the plastic compaction, so that a progressive build up of pore pressure is observed prior to failure. The failure mechanism involves excessive deformation which develops at a constant axial stress level.

Finally, the simulations shown in Figure 3.16c have been carried out assuming

$$G = 20\,\text{MPa}; \quad K = 40\,\text{MPa}; \quad \eta_f = 1.15; \quad \eta_c = 1.15; \quad A = 0.005$$

which represents a set of parameters typical of sand in a *very loose* state of compaction. Here $\eta_c = \eta_f$; as a result, the generation of significant positive pore pressures takes place and the effective stress path migrates towards the stress space origin. At some stage the material characteristic becomes unstable, leading to a complete liquefaction of the sample.

The undrained characteristics, as presented here, are fairly consistent with the experimental evidence, Figure 3.5. Overall, the deviatoric hardening framework, in spite of its apparent limitations, can depict the basic trends in the mechanical response for a broad range of initial degrees of compaction.

3.3.3 Generalization and specification of the constitutive matrix

The generalization of the functional forms of the loading and plastic potential surfaces can be carried out by following the procedure outlined in Section 3.2.3. Introducing representation (3.44) in the expressions (3.61) and (3.64), respectively, yields

$$f = \sqrt{3}\bar{\sigma} - \eta\sigma_m\, g(\theta) = 0; \quad \psi = \sqrt{3}\bar{\sigma} + \eta_c\, g(\theta)\sigma_m \ln\frac{\sigma_m}{\sigma_m^0} = 0 \tag{3.67}$$

In addition, the notion of the deviatoric plastic strain rates, as defined through eq.(3.4), can be generalized, so that the hardening function becomes

$$\eta = \eta(\varepsilon_q^p); \quad d\varepsilon_q^p = \frac{2}{\sqrt{3}}\sqrt{J_{2\dot{\varepsilon}}} \tag{3.68}$$

where $J_{2\dot{\varepsilon}}$ represents the second invariant of the deviatoric part of the plastic strain increment.

In order to derive the constitutive relation, standard plasticity procedure can now be employed. The consistency condition takes the form

$$f(\boldsymbol{\sigma}, \varepsilon_q^p) = 0; \quad df = \left(\frac{\partial f}{\partial \boldsymbol{\sigma}}\right)^T d\boldsymbol{\sigma} + \frac{\partial f}{\partial \varepsilon_q^p} d\varepsilon_q^p = 0 \tag{3.69}$$

Assuming the additivity postulate, the response in the elastic range is defined as

$$d\boldsymbol{\sigma} = [D^e](d\boldsymbol{\varepsilon} - d\boldsymbol{\varepsilon}^p); \quad d\boldsymbol{\varepsilon} = d\boldsymbol{\varepsilon}^e + d\boldsymbol{\varepsilon}^p \tag{3.70}$$

whereas the plastic strain increments satisfy

$$d\boldsymbol{\varepsilon}^p = d\lambda \frac{\partial \psi}{\partial \boldsymbol{\sigma}}; \quad d\varepsilon_q^p = \frac{2}{\sqrt{3}} \sqrt{J_{2\psi}}\, d\lambda \tag{3.71}$$

where $J_{2\psi}$ is the second invariant of the deviatoric part of the gradient of the potential function. Thus, utilizing eqs. (3.69)–(3.71), the constitutive relation can be written in the form

$$d\boldsymbol{\sigma} = [D^{ep}]\, d\boldsymbol{\varepsilon}, \quad [D^{ep}] = [D^e] - \frac{1}{H}\left([D^e] \frac{\partial \psi}{\partial \boldsymbol{\sigma}} \left(\frac{\partial f}{\partial \boldsymbol{\sigma}}\right)^T [D^e]\right) \tag{3.72}$$

where

$$H = H_e + H_p; \quad H_e = \left(\frac{\partial f}{\partial \boldsymbol{\sigma}}\right)^T [D^e] \frac{\partial \psi}{\partial \boldsymbol{\sigma}}; \quad H_p = -\frac{2}{\sqrt{3}} \sqrt{J_{2\psi}} \frac{\partial f}{\partial \boldsymbol{\varepsilon}_q^p} \tag{3.73}$$

The above expression completes the general mathematical formulation of the model.

3.4 COMBINED VOLUMETRIC-DEVIATORIC HARDENING

In this section, a different conceptual framework is reviewed which combines the notions of volumetric and deviatoric hardening. The formulation is primarily intended as an extension of the Critical State model to incorporate the dilatancy effect [27,28]. The latter allows for broadening the range of applicability to include granular materials in a dense state of compaction.

The mathematical formulation follows the general framework outlined in previous Sections 3.2 and 3.3. Using the $\{p, q\}$ invariants, the loading surface is assumed in the form

$$f(p, q, \zeta) = 0 \tag{3.74}$$

where

$$d\zeta = \vartheta d\varepsilon_q^p + de^p = \vartheta d\varepsilon_q^p - (1 + e_0) d\varepsilon_v^p \tag{3.75}$$

and ϑ is a material constant. Apparently, for $\vartheta = 0$, the classical volumetric hardening framework is recovered.

The approach typically involves an associated flow rule (3.23), so that

$$d\zeta = d\lambda \left[\vartheta \frac{\partial f}{\partial q} - (1 + e_0) \frac{\partial f}{\partial p}\right] \tag{3.76}$$

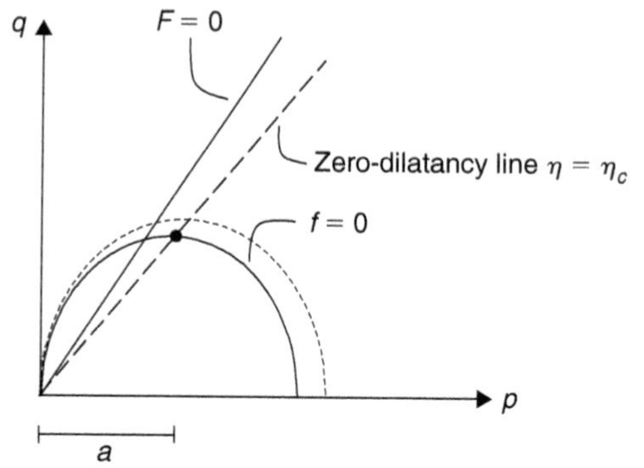

Figure 3.17 Schematic representation of the framework of combined volumetric-deviatoric hardening

The constitutive relation is derived by invoking the consistency condition

$$df = \frac{\partial f}{\partial p}dp + \frac{\partial f}{\partial q}dq + \frac{\partial f}{\partial \zeta}d\zeta = 0 \tag{3.77}$$

For a *stress-controlled* program, substitution of eq.(3.76) in (3.77) yields

$$d\lambda = \frac{1}{H_p}\left(\frac{\partial f}{\partial p}dp + \frac{\partial f}{\partial q}dq\right); \quad H_p = -\frac{\partial f}{\partial \zeta}\left[\vartheta\frac{\partial f}{\partial q} - (1+e_0)\frac{\partial f}{\partial p}\right] \tag{3.78}$$

so that

$$d\varepsilon_v^p = \frac{1}{H_p}\left\{\left(\frac{\partial f}{\partial p}\right)^2 dp + \frac{\partial f}{\partial p}\frac{\partial f}{\partial q}dq\right\}; \quad d\varepsilon_q^p = \frac{1}{H_p}\left\{\frac{\partial f}{\partial p}\frac{\partial f}{\partial q}dp + \left(\frac{\partial f}{\partial q}\right)^2 dq\right\} \tag{3.79}$$

The above representation leads to the constitutive relation that is analogous to that defined in (3.30). Apparently, for a *strain-controlled* program, the functional form (3.28) is recovered, with H_p as specified in (3.78).

Consider now the Critical State framework, as presented in Section 3.2, which is enriched by incorporating the combined volumetric-deviatoric hardening parameter ζ. The loading surface, as depicted in Figure 3.17, assumes the form

$$f(p,q,\zeta) = (p-a)^2 + \left(\frac{q}{\eta_c}\right)^2 - a^2 = 0; \quad a = a(\zeta) \tag{3.80}$$

Note that, the condition of no plastic volume change, i.e. $d\varepsilon_v^p = 0$, requires $p = a$; $q = \eta_c p$; so that the parameter η_c defines the location of *zero-dilatancy line*. This definition is, in fact, analogous to that employed in the context of deviatoric hardening, eq.(3.65). At the same time, the conditions at failure are governed by the requirement of $H_p = 0$. Assuming that the failure criterion is consistent with Mohr-Coulomb representation, i.e. $\eta = q/p = \eta_f$, the following expression is obtained

$$H_p = 0 \quad \Rightarrow \quad \vartheta = \frac{1}{2}(1 + e_0)\eta_f\left[\left(\frac{\eta_c}{\eta_f}\right)^2 - 1\right] \tag{3.81}$$

The above equation defines the parameter ϑ, which explicitly depends on the ratio η_c/η_f. Clearly, $\eta_c = \eta_f$ implies $\vartheta = 0$, so that the classical Critical State framework is recovered, in which the conditions at failure are coupled with no volume change. The material function $a = a(\zeta)$ is typically assumed in the functional form similar to eq.(3.37), i.e.

$$a = a_0 \exp\left(\frac{-\zeta}{\lambda - \kappa}\right) \tag{3.82}$$

which requires that the elastic bulk modulus be proportional to confining pressure, viz. eq.(3.38).

The formulation as outlined above incorporates one additional parameter, i.e. η_c, on top of the standard Critical State constants G, κ, ϕ and λ. Again, all constants can be defined from standard 'triaxial' tests involving hydrostatic compression (λ, κ) as well as axial compression (G, ϕ, η_c).

The performance of the framework is illustrated in Figures 3.18 and 3.19. The material parameters selected here are identical to those employed in Section 3.2.2., i.e.

$$G = 30\,\text{MPa}, \quad \kappa = 0.02, \quad \eta_f = 1.2, \quad \lambda = 0.13, \quad e_0 = 0.9$$

while $\eta_c = 1.0$.

The behaviour under *drained* conditions is qualitatively similar to that depicted in Figure 3.15. As the axial stress increases, a progressive transition from compaction to dilatancy occurs, Figure 3.18b. Clearly, for $\eta_c = \eta_f$, the prediction for Critical State model, Figure 3.11, is recovered.

Under *undrained* constraint, Figure 3.19, an initial build up of pore pressure is observed. When the transition to plastic dilation occurs, i.e. $\eta = \eta_c$, a generation of negative excess of pore pressure takes place and the effective stress trajectory moves towards the failure envelope. Again, the results for $\eta_c = \eta_f$ are consistent with those shown in Figure 3.12. Overall, the results as depicted in Figures 3.18 and 3.19 clearly indicate that by incorporating a combined volumetric-deviatoric hardening, the range of applicability of the Critical State framework can be extended to include the granular materials in a dense/medium dense state of compaction.

Finally, the framework outlined here can be generalized by following the procedure analogous to that in Sections 3.2.3. and 3.3.3. The functional form of the loading surface is now similar to that in eq.(3.45), i.e.

$$f = (\sigma_m - a)^2 + \left(\frac{\sqrt{3}\bar{\sigma}}{\eta_c g(\theta)}\right)^2 - a^2 = 0 \tag{3.83}$$

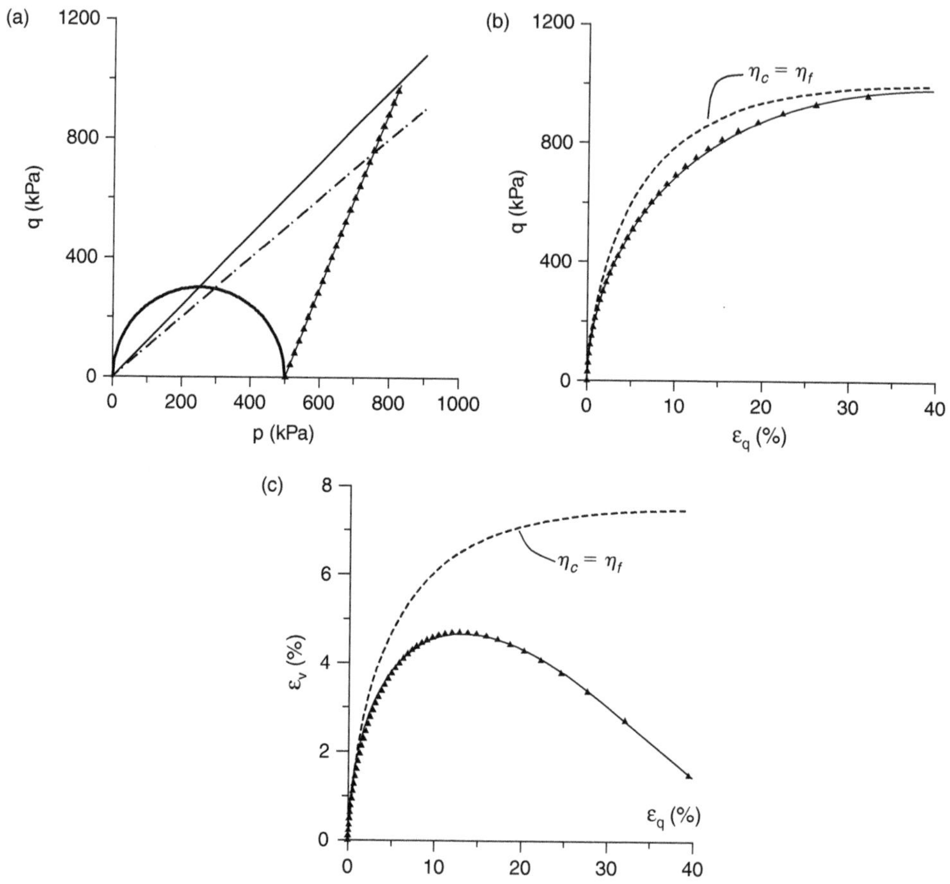

Figure 3.18 Combined volumetric-deviatoric hardening; numerical predictions for a drained axial compression test

where $a = a(\zeta)$. The evolution law for the hardening parameter ζ, eq.(3.75), becomes

$$d\zeta = d\lambda \left[\vartheta \frac{2}{\sqrt{3}} \sqrt{J_{2\psi}} - (1 + e_0)\, \boldsymbol{\delta}^T \frac{\partial f}{\partial \boldsymbol{\sigma}} \right] \tag{3.84}$$

where $J_{2\psi}$ is the second invariant of the deviatoric part of the plastic potential gradient, eq.(3.71), which in this case is the same as that of the loading function f.

The constitutive relation takes the form analogous to that specified in (3.51), i.e.

$$d\boldsymbol{\sigma} = [D^{ep}]d\boldsymbol{\varepsilon}, [D^{ep}] = [D^e] - \frac{1}{H}\left([D^e]\frac{\partial f}{\partial \boldsymbol{\sigma}}\left(\frac{\partial f}{\partial \boldsymbol{\sigma}}\right)^T [D^e]\right) \tag{3.85}$$

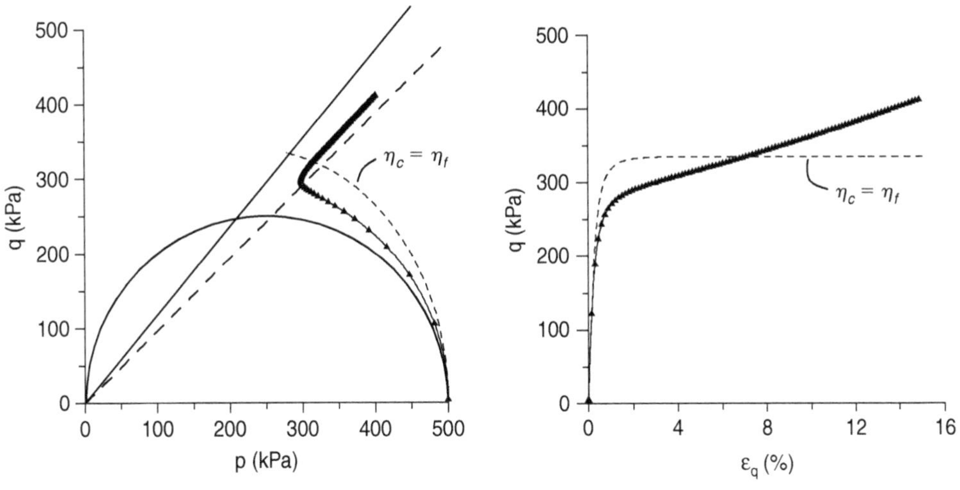

Figure 3.19 Combined volumetric-deviatoric hardening; numerical simulations of an undrained compression test

where

$$H_e = \left(\frac{\partial f}{\partial \boldsymbol{\sigma}}\right)^T [D^e]\frac{\partial f}{\partial \boldsymbol{\sigma}}; \quad H_p = -\frac{\partial f}{\partial \zeta}\left[\vartheta\frac{2}{\sqrt{3}}\sqrt{J_{2\psi}} - (1+e_0)\,\boldsymbol{\delta}^T\frac{\partial f}{\partial \boldsymbol{\sigma}}\right] \tag{3.86}$$

Again, $[D^{ep}]$ is a positive-definite matrix, provided $H_p > 0$.

3.5 SPECIFICATION OF CONSTITUTIVE MATRIX UNDER UNDRAINED CONDITIONS

As a conclusion to this chapter, it may be instructive to examine the form of the constitutive relation in case of undrained constraint. Let us restrict ourselves to a class of geomaterials for which the Terzaghi's effective stress principle applies, e.g. cohesionless soils. In this case, the loading function as well as conditions at failure both depend on the effective stress, so that

$$f(\boldsymbol{\sigma}, \kappa) = 0 \quad \Rightarrow \quad d\boldsymbol{\sigma} = [D^{ep}]\, d\boldsymbol{\varepsilon} \tag{3.87}$$

The Terzaghi's effective stress principle reads

$$d\hat{\boldsymbol{\sigma}} = d\boldsymbol{\sigma} + dp_f\boldsymbol{\delta} \tag{3.88}$$

where $\hat{\boldsymbol{\sigma}}$ is the total stress and $p_f = -\boldsymbol{\delta}^T \boldsymbol{\sigma}_f$ is the fluid pressure.

The undrained conditions imply that the fluid (usually water) cannot escape from voids, so that the overall macroscopic deformation must be consistent with that of the fluid. Thus, the kinematic constraint of undrained deformation requires

$$d\boldsymbol{\varepsilon}_f = n^{-1}\, d\boldsymbol{\varepsilon}_v \tag{3.89}$$

where ε_f is the volumetric strain in the fluid and n is the porosity. Assume that, in the range of pore pressures considered, the fluid is linearly compressible, with average bulk modulus of K_f. Substituting eqs.(3.87) and (3.88) in (3.89) results, after some algebraic transformations, in

$$d\hat{\boldsymbol{\sigma}} = [\hat{D}]\, d\boldsymbol{\varepsilon}; \quad [\hat{D}] = [D^{ep}] + \boldsymbol{\delta}\frac{K_f}{n}\boldsymbol{\delta}^T \tag{3.90}$$

The problem can also be formulated in terms of compliance operator. In this case

$$d\boldsymbol{\varepsilon} = [C^{ep}]d\boldsymbol{\sigma}; \quad d\boldsymbol{\sigma} = d\hat{\boldsymbol{\sigma}} - dp_f\boldsymbol{\delta} \tag{3.91}$$

where $[C^{ep}] = [D^{ep}]^{-1}$ is the elastoplastic compliance. The kinematic constraint of undrained deformation reads

$$\boldsymbol{\delta}^T d\boldsymbol{\varepsilon} = \boldsymbol{\delta}^T[C^{ep}]d\boldsymbol{\sigma} - \boldsymbol{\delta}^T[C^{ep}]\boldsymbol{\delta}dp_f = \frac{n}{K_f}dp_f \tag{3.92}$$

which results in

$$dp_f = \frac{\boldsymbol{\delta}^T[C^{ep}]\, d\boldsymbol{\sigma}}{\boldsymbol{\delta}^T[C^{ep}]\boldsymbol{\delta} + \dfrac{n}{K_f}} \tag{3.93}$$

Thus,

$$d\boldsymbol{\varepsilon} = [\hat{C}]d\boldsymbol{\sigma}; \quad [\hat{C}] = [C^{ep}] - \frac{[C^{ep}]\boldsymbol{\delta}\boldsymbol{\delta}^T[C^{ep}]}{\boldsymbol{\delta}^T[C^{ep}]\boldsymbol{\delta} + \dfrac{n}{K_f}} \tag{3.94}$$

It is noted that the compressibility of the fluid is, in general, much larger than that of the soil skeleton, i.e. $K_f >> K$, so that according to eq.(3.92) $d\varepsilon_v \to 0$.

Chapter 4

Combined isotropic-kinematic hardening rules

The analysis in the preceding chapter has been primarily focused on description of monotonic loading histories that involve generation of irreversible/plastic deformation. In this context, the isotropic strain-hardening idealization is quite efficient in terms of reproducing the basic trends in the mechanical response of geomaterials. One of the limitations though is the fact that for all trajectories penetrating the domain enclosed by the current loading surface the behaviour is said to be elastic. This, in general, is acceptable provided the loading is restricted to histories that do not involve repeated action of loads, such as those associated with seismic excitation, sea wave action, etc. In fact, during the reverse loading a significant plastic deformation may develop that will affect the material response. This is particularly evident in saturated soils, where such diversified effects as liquefaction and cyclic mobility occur.

The response of geomaterials subjected to cyclic loading is typically described by invoking *anisotropic hardening* rules, which involve a combination of isotropic and kinematic hardening. The most common approach is the so-called *bounding surface plasticity* formulation. This framework was developed in the mid 1970's and the early references include the works of Mroz and co-workers [29,30], Dafalias and Popov [31,32] as well as Krieg [33]. In this chapter, two basic approaches are examined. The first one represents an extension of volumetric hardening (Critical State) framework, while the second one deals with the deviatoric hardening. Here, both these approaches are reformulated within the framework of bounding surface plasticity. Combined isotropic-kinematic hardening rules are introduced to account for the generation of plastic deformation during loading histories experiencing stress reversals. The performance of both these frameworks is illustrated by some numerical examples.

4.1 BOUNDING SURFACE PLASTICITY; VOLUMETRIC HARDENING FRAMEWORK

The basic notion incorporated here is that of a *bounding* or *consolidation* surface, which is defined by invoking the volumetric hardening rule as described in Section 3.2. This surface is constituted by an initial active loading history such as, for example, the in-situ consolidation process. Thus, as long as the stress point remains on the bounding surface, the material undergoes isotropic hardening and the formulation of the problem is consistent with that given in Sections 3.2.1 and 3.2.3. Now, for all stress reversal histories, the elastic behaviour is restricted to the domain enclosed by the *yield* surface, while beyond this region the plastic deformation is said to occur. The problem is formulated by invoking a combined isotropic-kinematic hardening rule. In particular, a translation rule for the yield surface and an interpolation rule for the field of hardening moduli are postulated to define the material response. The interpolation rule relates the hardening modulus to the distance between the yield and the bounding surfaces. When the stress trajectory reaches the bounding surface again, the memory of the past stress reversal event is erased and the formulation reverts back to the isotropic hardening framework.

4.1.1 Formulation in the 'triaxial' {p,q} space

Consider first an active loading history, such as initial consolidation under hydrostatic or non-hydrostatic conditions. In this case, the deformation process is described in terms of evolution of the loading surface $f(p,q,e^p)=0$, eq.(3.33). Thus

$$\begin{aligned} f &= (p-a)^2 + (q/\eta_f)^2 - a^2 = 0 \quad (for\ q > 0) \\ f &= (p-a)^2 + (q/\tilde{\eta}_f)^2 - a^2 = 0 \quad (for\ q < 0) \end{aligned} \tag{4.1}$$

where $a=a(e^p)$. Note again that $f=0$ is, in fact, asymmetric; i.e. it is composed of two distinct semi-ellipses in compression and extension domains, respectively.

The plastic strain is defined by invoking an associated flow rule, eq.(3.23), and imposing the consistency condition $df=0$, eq.(3.24). According to representation (3.29)

$$d\varepsilon_v^p = \frac{1}{H_p}\left(\frac{\partial f}{\partial p}dp + \frac{\partial f}{\partial q}dq\right)\frac{\partial f}{\partial p}; \qquad d\varepsilon_q^p = \frac{1}{H_p}\left(\frac{\partial f}{\partial p}dp + \frac{\partial f}{\partial q}dq\right)\frac{\partial f}{\partial q} \tag{4.2}$$

where the hardening modulus H_p is specified in eq.(3.37), i.e.

$$H_p = (1+e_0)\frac{4\,a\,p(p-a)}{\lambda - \kappa} \tag{4.3}$$

Thus, the formulation above is that of classical Critical State theory and it is valid as long as $f=0$, i.e. the stress point remains on the current loading surface.

Consider now a stress reversal history, for which $f<0$. In this case, the stress path penetrates the region inside the loading surface, which is now referred to as the *bounding surface*. For such trajectories, the deformation process is described in terms of evolution of the *yield surface* $f_0=0$, which undergoes translation within the domain enclosed by the bounding surface. Referring to Figure 4.1a, the yield surface $f_0=0$ is initially tangential to the bounding surface at the stress reversal point R. For the subsequent history, say the path RSS', the behaviour is elastic for RS (since $f_0<0$) and becomes elastoplastic for SS' triggering the translation of $f_0=0$ as the stress point proceeds along its trajectory.

The equation of the yield surface takes the form

$$f_0 = (p-\alpha_p)^2 + \left(\frac{q-\alpha_q}{\eta_f}\right)^2 - a_0^2 = 0; \quad for\ q-\alpha_q > 0 \tag{4.4}$$

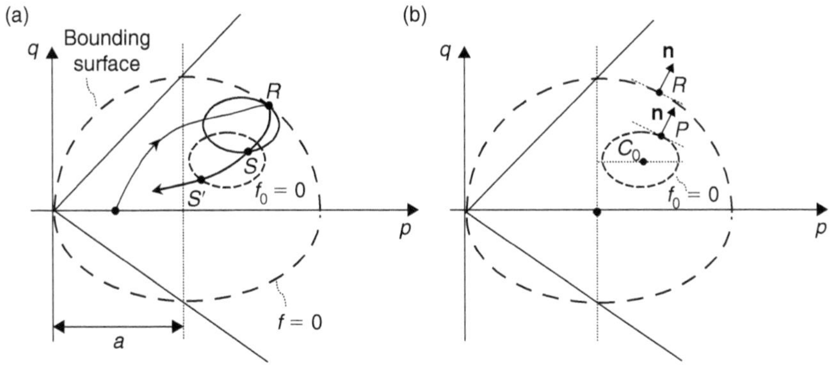

Figure 4.1 Volumetric hardening framework; bounding and yield surfaces in 'triaxial' {p,q} space

where a_0 is the size parameter while α_p, α_q are the coordinates that specify the current position of $f_0 = 0$, viz point C_0 (Figure 4.1b). Here, again, the functional form of $f_0 = 0$ for $q - \alpha_q < 0$ is obtained by replacing η_f by $\tilde{\eta}_f$, similar to eq.(4.1).

The description of plastic deformation involves an anisotropic hardening rule that requires specification of the field of hardening moduli together with the kinematics of the yield surface. With this in mind, it is convenient to write the flow rule in a normalized form that is consistent with representation (4.2), i.e.

$$d\varepsilon_v^p = \frac{1}{K_p}(n_p dp + n_q dq)n_p; \quad d\varepsilon_q^p = \frac{1}{K_p}(n_p dp + n_q dq)n_q \tag{4.5}$$

Here,

$$n_p = \frac{\partial f_0}{\partial p}\left[\left(\frac{\partial f_0}{\partial p}\right)^2 + \left(\frac{\partial f_0}{\partial q}\right)^2\right]^{-1/2}; \quad n_q = \frac{\partial f_0}{\partial q}\left[\left(\frac{\partial f_0}{\partial p}\right)^2 + \left(\frac{\partial f_0}{\partial q}\right)^2\right]^{-1/2} \tag{4.6}$$

are the components of the unit vector along the gradient of the yield function f_0 and K_p is the normalized hardening modulus. The latter is obtained by imposing an interpolation rule that relates the value of K_p to the distance between the yield and the bounding surface. In particular, the following function may be assumed

$$K_p = K_R + K_0 \left({}^{\delta}/_{\delta_0}\right)^{\gamma} \tag{4.7}$$

In this expression, K_R is the normalized hardening modulus evaluated on the bounding surface at the so-called *conjugate* point. Referring to Figure 4.1b, if the current stress point is at P its conjugate (i.e. point R) is defined as a point on the bounding surface at which the direction of the gradient vector is the same as that at P on $f_0 = 0$. Thus,

$$K_R = H_p \left[\left(\frac{\partial f_0}{\partial p} \right)^2 + \left(\frac{\partial f_0}{\partial q} \right)^2 \right]^{-1} \tag{4.8}$$

where H_p is given by eq.(4.3) and it is evaluated at point R. Furthermore, δ is the current distance between the stress and the conjugate points, $\delta = PR$; δ_0 denotes the maximum distance, $\delta_0 = 2(a - a_0)$, and γ is a material constant. Note that $\delta \to 0 \Rightarrow K_p \to K_R$ while for $\delta \to \delta_0$ there is $K_p \to K_0$. The latter holds since K_0, which defines the initial value of the modulus, is assumed to satisfy $K_0 >> K_R$.

Consider now the kinematics of the yield surface. For this purpose, it is convenient to employ the matrix notation, i.e. define the current stress state as $\boldsymbol{\sigma} = \boldsymbol{\sigma}^P = \{p, q\}^T$ and similarly, refer to the location of the yield surface (point C_0) as $\boldsymbol{\alpha} = \{\alpha_p, \alpha_q\}^T$. In view of geometric similarities of both surfaces, Figure 4.1b, the following relations hold

$$\boldsymbol{\sigma}^R - \boldsymbol{\alpha}^R = \frac{a}{a_0}(\boldsymbol{\sigma}^P - \boldsymbol{\alpha}); \quad \boldsymbol{\beta} = \boldsymbol{\sigma}^R - \boldsymbol{\sigma}^P = \frac{1}{a_0}[(a - a_0)\boldsymbol{\sigma}^P - (a\boldsymbol{\alpha} - a_0\boldsymbol{\alpha}^R)] \tag{4.9}$$

where $\boldsymbol{\alpha}^R = \{a, 0\}^T$.

Assume now that both surfaces do not intersect, but instead engage each other along a common normal. In this case, the translation rule may be formulated by postulating that relative motion of point P with respect to R is directed along $\boldsymbol{\beta}$; in other words $d\boldsymbol{\sigma}^P - d\boldsymbol{\sigma}^R = \boldsymbol{\beta} d\mu$, where $d\mu$ is a scalar factor. Considering that both surfaces can translate and expand, the following incremental relations can be established for the change in position of P and R (cf. [29], [34])

$$d\boldsymbol{\sigma}^R = d\boldsymbol{\alpha}^R + \frac{da}{a}(\boldsymbol{\sigma}^R - \boldsymbol{\alpha}^R); \quad d\boldsymbol{\sigma}^P = d\boldsymbol{\alpha} + \frac{da_0}{a_0}(\boldsymbol{\sigma}^P - \boldsymbol{\alpha}) \tag{4.10}$$

Thus, substituting (4.9) and (4.10) in the proposed translation rule results in

$$d\boldsymbol{\alpha} = \boldsymbol{\beta}\, d\mu + (\boldsymbol{\sigma}^P - \boldsymbol{\alpha})\frac{da - da_0}{a_0} + d\boldsymbol{\alpha}^R \tag{4.11}$$

which defines the kinematics of the yield surface.

Note that the scalar parameter $d\mu$ can be obtained from the consistency condition $df_0 = 0$, i.e.

$$df_0 = \left(\frac{\partial f_0}{\partial \boldsymbol{\sigma}} \right)^T d\boldsymbol{\sigma} + \left(\frac{\partial f_0}{\partial \boldsymbol{\alpha}} \right)^T d\boldsymbol{\alpha} + \left(\frac{\partial f_0}{\partial a_0} \right) da_0 = 0 \tag{4.12}$$

Substituting eq.(4.11) and noting that $\boldsymbol{\sigma}^P = \boldsymbol{\sigma}$ while $\partial f_0/\partial\boldsymbol{\alpha} = -\partial f_0/\partial\boldsymbol{\sigma}$ yields

$$d\mu = \frac{\left(\frac{\partial f_0}{\partial \boldsymbol{\sigma}}\right)^T \left[d\boldsymbol{\sigma} - (\boldsymbol{\sigma} - \boldsymbol{\alpha})\frac{da - da_0}{a_0} - d\boldsymbol{\alpha}^R\right]}{\left(\frac{\partial f_0}{\partial \boldsymbol{\sigma}}\right)^T \boldsymbol{\beta}} \tag{4.13}$$

The kinematic hardening formulation given here is the simplest, as it employs two distinct surfaces only. In general, various extensions of this framework have been proposed in the literature. Those incorporate, for example, a set of nesting surfaces within the domain contained between $f = 0$ and $f_0 = 0$ (cf.[29],[34]) or an infinite number of active loading/stress reversal surfaces generated during repeated stress reversals [35,36]. In both these approaches, the functional form of the interpolation, eq.(4.7), and translation rules, eq.(4.11), is actually the same while the formulation engages more complex memory rules.

The two-surface approach outlined here, in spite of its simplicity, is quite efficient in terms of depicting the basic trends in mechanical response, as demonstrated later in this chapter. Note that in numerical implementation, the translation rule is often simplified by neglecting the expansion of both surfaces (cf. [29],[34]) i.e. by postulating

$$d\boldsymbol{\alpha} = \boldsymbol{\beta}\, d\mu \Rightarrow d\mu = \frac{\left(\frac{\partial f_0}{\partial \boldsymbol{\sigma}}\right)^T d\boldsymbol{\sigma}}{\left(\frac{\partial f_0}{\partial \boldsymbol{\sigma}}\right)^T \boldsymbol{\beta}} \tag{4.14}$$

Or, in explicit terms

$$\begin{aligned} d\alpha_p &= d\mu(p^R - p); \quad d\alpha_q = d\mu(q^R - q); \\ d\mu &= \left(\frac{\partial f_0}{\partial p}dp + \frac{\partial f_0}{\partial q}dq\right) \Big/ \left(\frac{\partial f_0}{\partial p}(p^R - p) + \frac{\partial f_0}{\partial q}(q^R - q)\right) \end{aligned} \tag{4.15}$$

Finally, invoking the flow rule (4.5) together with the additivity postulate, the constitutive relation assumes the form that is consistent with representation (3.30) in the previous chapter, i.e.

$$\begin{Bmatrix} d\varepsilon_v \\ d\varepsilon_q \end{Bmatrix} = \begin{bmatrix} \left(\frac{1}{K} + \frac{1}{K_p}n_p n_p\right) & \left(\frac{1}{K_p}n_p n_q\right) \\ \left(\frac{1}{K_p}n_p n_q\right) & \left(\frac{1}{3G} + \frac{1}{K_p}n_q n_q\right) \end{bmatrix} \begin{Bmatrix} dp \\ dq \end{Bmatrix} \tag{4.16}$$

subject to the interpolation, eq.(4.7), and translation rule, eq.(4.11) or eq.(4.14).

4.1.2 Comments on the performance

Let us examine now some basic trends in the mechanical response as predicted by the two-surface model. For this purpose choose the same material parameters as those employed in Section 3.2.2, i.e.

$$G = 30\,\text{MPa}, \quad \phi = 30°, \quad \lambda = 0.13, \quad \kappa = 0.02, \quad e_0 = 0.9$$

which may be considered as representative of a normally consolidated clay. Let us analyze the behaviour of a sample tested in axial compression at the confinement of 500 kPa. More specifically, let us focus on undrained conditions, analogous to those depicted in Figure 3.12, and examine the response under both stress and strain-controlled cyclic loading conditions.

Implementation of the anisotropic hardening rule requires the identification of constants K_0, γ that appear in the interpolation function (4.7), as well as the information on the rate of expansion of the yield surface. The latter is typically assumed to be proportional to that of the bounding surface (i.e., $a_0/a = \text{const.}$) while the parameters in the interpolation function are specified by trial and error. Given that the focus here is on a qualitative performance of the framework assume, for simplicity, that the size of the yield surface is fixed, say $a_0 = 20$ kPa (i.e., $a_0 << a$) and take some typical values of $K_0 = 10 \times 10^6$ kPa, $\gamma = 3$.

The results of numerical simulations are shown in Figures 4.2–4.4. Figure 4.2 presents the results of a two-way strain-controlled undrained cyclic test with the axial strain amplitude of $\varepsilon_q = -\varepsilon_1 = \pm 0.5\%$. Figure 4.2a shows the effective stress trajectories, while Figure 4.2b presents the corresponding strain characteristic. Note that in the course of cyclic loading the effective stress path progressively migrates towards the origin. The resultant stress amplitude remains nearly constant for the consecutive

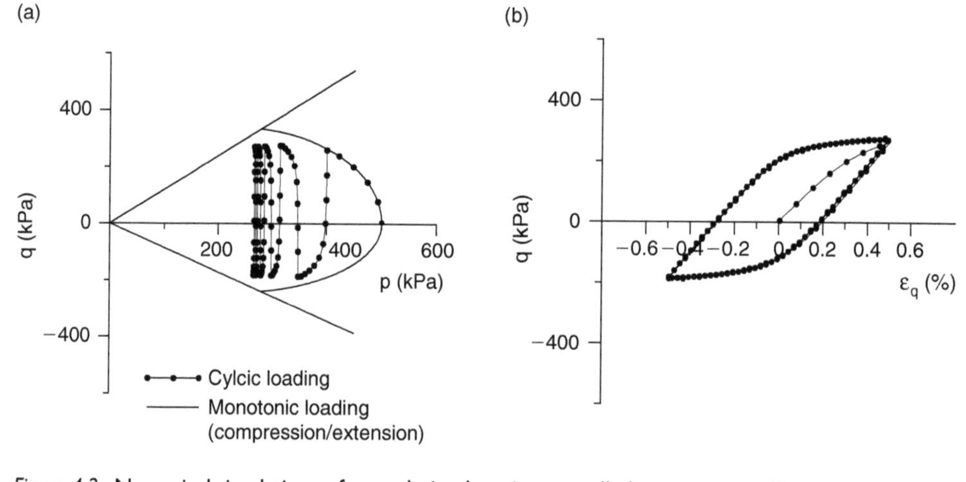

Figure 4.2 Numerical simulations of an undrained strain-controlled test on normally consolidated clay (10 strain-reversal cycles with amplitude ±0.5%); (a) effective stress paths, (b) deformation characteristic

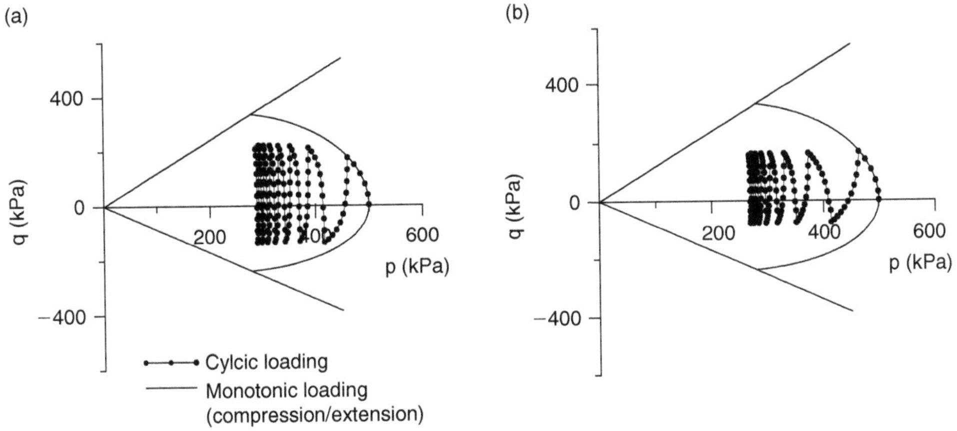

Figure 4.3 Numerical simulations of an undrained strain-controlled test on normally consolidated clay (15 strain-reversal cycles with amplitude ±0.25%); effective stress trajectories for (a) $\gamma = 3$, (b) $\gamma = 5$

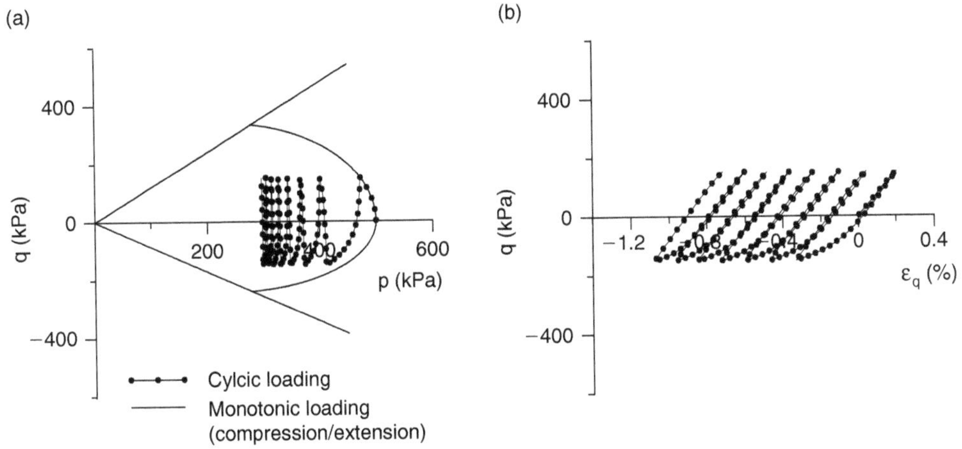

Figure 4.4 Numerical simulations of an undrained stress-controlled test on normally consolidated clay (15 stress-reversal cycles with amplitude ±150 kPa); (a) effective stress paths, (b) deformation characteristic

cycles. After a number of strain reversals, a stationary state is reached in which the strain hysteresis repeats itself in each cycle.

Figure 4.3 shows the results of a parametric study that illustrates the influence of strain amplitude as well as the value of parameter γ in the interpolation function (4.7). The simulations shown here correspond to strain amplitude of ±0.25%, while $\gamma = 3$ and $\gamma = 5$, respectively. Here, the trends are similar to those depicted in Figure 4.2. Decrease in the strain amplitude results in a corresponding decrease in the resultant stress amplitude. Note that the stress amplitudes in extension domain ($q < 0$) are lower than those in compression ($q > 0$). Figure 4.3b gives the results for the same strain amplitude and $\gamma = 5$. Comparing the two sets of results in Figure 4.3, it is evident that an increase in the value of γ leads to a further decrease in the stress amplitude and a more significant reduction in the effective pressure.

Figure 4.4 shows the results of simulations that correspond to a two-way stress-controlled program. Here, the stress amplitudes are fixed at ±150 kPa in compression and extension domains, respectively. In this case, the so-called *cyclic mobility* conditions are reached. The effective stress path stabilizes, no liquefaction is achieved, and the deformation accumulates in each consecutive cycle.

The trends as depicted here are fairly consistent with the experimental data on cyclic response of saturated clays. The reader is referred at this point to Chapter 8, which provides a more general discussion on the mechanical behaviour of geomaterials.

4.1.3 Generalization and specification of the constitutive matrix

In order to establish the functional form of the yield surface, consider first the basic invariants of the stress tensor $\underset{\sim}{\sigma}_{ij} = \sigma_{ij} - \alpha_{ij}$, where the operator α_{ij} defines the location of the yield surface in the general stress space. Thus,

$$\underset{\sim}{I}_1 = \underset{\sim}{\sigma}_{ii} = \sigma_{ii} - \alpha_{ii}; \quad \underset{\sim}{J}_2 = \frac{1}{2}\underset{\sim}{s}_{ij}\underset{\sim}{s}_{ij}; \quad \underset{\sim}{J}_3 = \frac{1}{2}\underset{\sim}{s}_{ij}\underset{\sim}{s}_{jk}s_{ki} \tag{4.17}$$

where

$$\underset{\sim}{s}_{ij} = \underset{\sim}{\sigma}_{ij} - \frac{1}{3}\delta_{ij}\underset{\sim}{\sigma}_{kk} = \left(\sigma_{ij} - \frac{1}{3}\delta_{ij}\sigma_{kk}\right) - \left(\alpha_{ij} - \frac{1}{3}\delta_{ij}\alpha_{kk}\right) = s_{ij} - \bar{\alpha}_{ij} \tag{4.18}$$

The modified stress invariants, as defined in representation (2.13) in Chapter 2, become

$$\underset{\sim}{\sigma}_m = -\frac{1}{3}\underset{\sim}{I}_1; \quad \underset{\sim}{\bar{\sigma}} = (\underset{\sim}{J}_2)^{½}; \quad \underset{\sim}{\theta} = \frac{1}{3}\sin^{-1}\left(\frac{-3\sqrt{3}\,\underset{\sim}{J}_3}{2\;\underset{\sim}{\bar{\sigma}}^3}\right) \wedge -\frac{\pi}{6} \le \underset{\sim}{\theta} \le \frac{\pi}{6} \tag{4.19}$$

Note that under the 'triaxial' conditions, i.e. $\sigma_2 = \sigma_3$, $\alpha_2 = \alpha_3$, the above expressions reduce to

$$\begin{aligned}
&\underset{\sim}{\sigma}_m = -\frac{1}{3}(\underset{\sim}{\sigma}_1 + 2\underset{\sim}{\sigma}_3) = p - \alpha_p; \quad \underset{\sim}{\bar{\sigma}} = \frac{1}{\sqrt{3}}|\underset{\sim}{\sigma}_3 - \underset{\sim}{\sigma}_1| = \frac{|q - \alpha_q|}{\sqrt{3}};\\
&\underset{\sim}{J}_3 = -\frac{2}{27}(\underset{\sim}{\sigma}_3 - \underset{\sim}{\sigma}_1)^3 \Rightarrow \underset{\sim}{\theta} = \pm\frac{\pi}{6}
\end{aligned} \tag{4.20}$$

The generalization of the functional form of the yield surface is now based on relations (4.20) combined with representation (2.25) of Chapter 2, i.e.

$$\underset{\sim}{\bar{\sigma}} = \underset{\sim}{\bar{\sigma}}_{|\theta=\frac{\pi}{6}}\, g(\underset{\sim}{\theta}) \quad \Rightarrow \quad |q - \alpha_q| = \sqrt{3}\underset{\sim}{\bar{\sigma}}/g(\underset{\sim}{\theta}); \quad p - \alpha_p = \underset{\sim}{\sigma}_m \tag{4.21}$$

Substituting the above equations in the expression for the yield surface (4.4) gives

$$f_0 = \underset{\sim}{\sigma}_m^2 + \left(\frac{\sqrt{3}\underset{\sim}{\bar{\sigma}}}{\eta_f g(\underset{\sim}{\theta})}\right)^2 - a_0^2 = 0 \tag{4.22}$$

For the stress states satisfying $f_0 = 0$, the flow rule (4.5) can be expressed in a general form

$$d\boldsymbol{\varepsilon}^p = \frac{1}{K_p}(\boldsymbol{n}^T d\boldsymbol{\sigma})\boldsymbol{n}; \quad \boldsymbol{n} = \frac{\partial f_0}{\partial \boldsymbol{\sigma}} / \left|\frac{\partial f_0}{\partial \boldsymbol{\sigma}}\right| \tag{4.23}$$

Here, K_p is defined in terms of interpolation function (4.7), i.e.

$$K_p = K_R + K_0\left({}^{\delta}/_{\delta_0}\right)^{\gamma}; \quad K_R = H_p / \left(\left|\frac{\partial f_0}{\partial \boldsymbol{\sigma}}\right|\right)^2; \quad \delta = \left|\boldsymbol{\sigma} - \boldsymbol{\sigma}^R\right| \tag{4.24}$$

Note that H_p, i.e. the hardening modulus at the conjugate point on the bounding surface, is defined according to eq.(3.50) in the preceding chapter, i.e.

$$H_p = \frac{\partial f}{\partial e^p}(1 + e_0)\,\mathrm{tr}\left(\frac{\partial f}{\partial \boldsymbol{\sigma}}\right) = (1 + e_0)\frac{2a\,\sigma_m}{\lambda - \kappa}\,\mathrm{tr}\left(\frac{\partial f}{\partial \boldsymbol{\sigma}}\right); \quad \mathrm{tr}\left(\frac{\partial f}{\partial \boldsymbol{\sigma}}\right) = \frac{\partial f}{\partial \sigma_{ii}} \tag{4.25}$$

The translation rule that defines the kinematics of the yield surface, has the same functional form as that of eq.(4.11), i.e.

$$d\boldsymbol{\alpha} = \boldsymbol{\beta} d\mu + (\boldsymbol{\sigma} - \boldsymbol{\alpha})\frac{da - da_0}{a_0} + d\boldsymbol{\alpha}^R \tag{4.26}$$

where $d\mu$ is evaluated from eq.(4.13). In this case, all operators are defined as vectors in a general six-dimensional stress space and $\boldsymbol{\alpha}^R = \{-a, -a, -a, 0, 0, 0\}^T$. Also note that the location of the conjugate stress point, as well as the value of $\boldsymbol{\beta}$, both can be directly obtained from relation (4.9).

The general form of the constitutive relation can be specified by invoking eq.(4.23) together with the additivity postulate. Thus,

$$d\boldsymbol{\varepsilon} = d\boldsymbol{\varepsilon}^e + d\boldsymbol{\varepsilon}^p = [C^{ep}]d\boldsymbol{\sigma}; \quad [C^{ep}] = [C^e] + \frac{1}{K_p}\boldsymbol{n}\,\boldsymbol{n}^T \tag{4.27}$$

where $[C^e] = [D^e]^{-1}$ is the elastic compliance operator.

Alternatively, for a strain-controlled history, the plastic strain rates may be defined as

$$d\boldsymbol{\varepsilon}^p = \frac{1}{\tilde{H}}(\boldsymbol{n}^T[D^e]\,d\boldsymbol{\varepsilon})\boldsymbol{n}; \quad \tilde{H} = (H_e + H_p) / \left(\left|\frac{\partial f_0}{\partial \boldsymbol{\sigma}}\right|\right)^2 \tag{4.28}$$

Thus, invoking the additivity of strain rate again, the constitutive relation becomes

$$d\boldsymbol{\sigma} = [D^e](d\boldsymbol{\varepsilon} - d\boldsymbol{\varepsilon}^p) = [D^{ep}]\,d\boldsymbol{\varepsilon}; \quad [D^{ep}] = [D^e] - \frac{1}{\tilde{H}}([D^e]\boldsymbol{n}\,\boldsymbol{n}^T[D^e]) \tag{4.29}$$

For the states on the bounding surface, the formulation can be directly obtained from the representation outlined above by setting $\boldsymbol{\alpha}=\boldsymbol{\alpha}^R$ and $a_0=a$. The modified stress invariants (4.19) become now

$$\underset{\sim}{\sigma}_m = -\frac{1}{3}(\sigma_{ii}+3a) = \sigma_m - a; \quad \underset{\sim}{\overline{\sigma}} = \overline{\sigma} = (J_2)^{1/2}; \quad \underset{\sim}{\theta} = \theta \tag{4.30}$$

so that

$$f = (\sigma_m - a)^2 + \left(\frac{\sqrt{3}\overline{\sigma}}{\eta_f g(\theta)}\right)^2 - a^2 = 0 \tag{4.31}$$

which is identical to eq.(3.45). Also, for $\delta=0$, which is associated with $f=0$, there is $K_p=K_R$ and the constitutive relation (4.29) is, in fact, the same as that given by eq.(3.51) in Section 3.2.3.

Finally, note that the framework presented here can be employed within the context of combined volumetric-deviatoric hardening, as discussed earlier in Section 3.4. In fact, the mathematical formulation of the problem remains the same except that $a=a(\zeta)$ rather than $a=a(e^p)$, where ζ is a hardening parameter whose evolution is defined as a linear combination of volumetric and deviatoric plastic strain rates, eq.(3.75).

4.2 BOUNDING SURFACE PLASTICITY; DEVIATORIC HARDENING FRAMEWORK

By examining the predictive abilities of the volumetric hardening framework, as outlined in Section 4.1, it is evident that it cannot describe a broad range of characteristics that are typical of saturated granular media in a relatively loose state of compaction. An example here is the liquefaction phenomenon that commonly occurs in loose/medium dense sand deposits subjected to cyclic excitation. A discussion on experimental studies involving cyclic loading that have provided a considerable insight into liquefaction and cyclic mobility mechanisms, is given later in Chapter 8.

In general, it appears that the framework of deviatoric hardening, as outlined in Section 3.3, is more suited for describing the response of granular soils. This is particularly evident when examining the results of numerical simulations given in Section 3.3.2. Therefore, in this section a bounding surface plasticity approach is presented, which focuses on extension of the classical deviatoric hardening model to the cyclic loading regime. The basic notions incorporated here are similar to those introduced in Section 4.1. In particular, the formulation for states on the bounding surface is consistent with that given in Sections 3.3.1 and 3.3.3. For trajectories within the domain enclosed by the bounding surface, the irreversible plastic deformation is said to occur and the problem is formulated by invoking an isotropic-kinematic hardening rule. Here, a two-surface framework is examined, which incorporates a non-associated flow rule. The translation rule for the yield surface and the interpolation function for the field of hardening moduli are postulated. The approach outlined here was developed in the early 1980's and the key references include the works reported in Refs.

[37, 38]. The presentation focuses, once again, on the 'triaxial' formulation first and is later extended to a general 3D stress state.

4.2.1 Formulation in the 'triaxial' {P,Q} space

Let us start with some preliminaries. The framework presented here incorporates a number of transformations in the effective stress space that impose constraints on the definition of stress measures p,q. In particular, the current position of the yield surface is represented in terms of vector $\boldsymbol{\alpha}=\{\alpha_P,\alpha_Q\}^T$, which is now defined as a *unit vector* directed along the axis of $f_0=0$. In order to ensure a proper generalization to 3D conditions, the components of $\boldsymbol{\alpha}$ should be defined as

$$\alpha_P=-\frac{1}{\sqrt{3}}(\alpha_1+2\alpha_3);\quad \alpha_Q=\sqrt{\frac{2}{3}}(\alpha_3-\alpha_1)\Rightarrow \alpha_P^2+\alpha_Q^2=\alpha_1^2+2\alpha_3^2=1 \tag{4.32}$$

This, in turn, requires a similar representation for the stress measures, i.e.

$$P=-\frac{1}{\sqrt{3}}(\sigma_1+2\sigma_3)=\sqrt{3}p;\quad Q=\sqrt{\frac{2}{3}}(\sigma_3-\sigma_1)=\sqrt{\frac{2}{3}}q \tag{4.33}$$

The latter ensures that the stress vector $\boldsymbol{\sigma}=\{P,Q\}^T$ follows the standard transformation rule of tensors of order one, i.e.

$$P'=P\alpha_P+Q\alpha_Q;\quad Q'=Q\alpha_P-P\alpha_Q \tag{4.34}$$

with α_P,α_Q being the direction cosines.

Given the representation (4.33), the strain rate measures that are compatible with stress parameters P, Q are now defined as

$$d\varepsilon_V=-\frac{1}{\sqrt{3}}(d\varepsilon_1+2d\varepsilon_3)=\frac{1}{\sqrt{3}}d\varepsilon_v;\quad d\varepsilon_Q=\sqrt{\frac{2}{3}}(d\varepsilon_3-d\varepsilon_1)=\sqrt{\frac{3}{2}}d\varepsilon_q \tag{4.35}$$

so that the rate of work becomes

$$dW=Pd\varepsilon_V+Qd\varepsilon_Q=\sigma_1 d\varepsilon_1+2\sigma_3\, d\varepsilon_3 \tag{4.36}$$

Given the definitions above, let us focus now on the formulation of the problem. For an initial active loading history, the deformation process is described in terms of evolution of the *bounding surface* $f=0$, eqs.(3.61)–(3.63). Incorporating the stress/strain measures viz. (4.33), (4.35), and considering both compression and extension regimes, yields

$$f = |Q| - g\eta P = 0; \qquad \eta = \upsilon_f \frac{\varepsilon_Q^p}{B + \varepsilon_Q^p}$$
$$g = \begin{cases} 1 & (for\ Q > 0) \\ \dfrac{(3 - \sin\phi)}{(3 + \sin\phi)} = \dfrac{2}{2 + \sqrt{2}\,\upsilon_f} = k & (for\ Q < 0) \end{cases} \tag{4.37}$$

Here, $f = 0$ is, once again, asymmetric, i.e. the branch in compression has a different slope than that in extension. Note that an alternative formulation is feasible here whereby, for $Q < 0$, g is expressed in terms of mobilized friction angle ϕ_m, rather than ϕ, so that g is taken as $g = 2/(2 + \sqrt{2}\eta) \neq const$. In this case for $\eta \to \upsilon_f$ there is $g \to k$, which is consistent with the Mohr-Coulomb failure criterion. On the other hand, $\eta \to 0 \Rightarrow g \to 1$, which means that at low deviatoric stress intensities $f = 0$ is nearly symmetric about the hydrostatic P-axis.

The plastic deformation is defined by invoking a non-associated flow rule, eq.(3.53). According to representation (3.60), there is

$$d\varepsilon_V^p = \frac{1}{H_p}\left(\frac{\partial f}{\partial P}dP + \frac{\partial f}{\partial Q}dQ\right)\frac{\partial \psi}{\partial P}; \quad d\varepsilon_q^p = \frac{1}{H_p}\left(\frac{\partial f}{\partial P}dP + \frac{\partial f}{\partial Q}dQ\right)\frac{\partial \psi}{\partial Q} \tag{4.38}$$

where ψ is the plastic potential function and H_p represents the hardening modulus. Both these functions are defined in accordance with eq.(3.64) and eq.(3.66), i.e.

$$\psi = |Q| + g\upsilon_c P \ln\left(\frac{P}{\overline{P}}\right) = 0; \quad H_p = -\frac{\partial f}{\partial \varepsilon_Q^p}\frac{\partial \psi}{\partial Q} = -\frac{(\upsilon_f - \eta)^2}{B\upsilon_f} gP \tag{4.39}$$

Thus, the formulation above is that of the standard deviatoric hardening, as discussed in section 3.3.1. Note that the modified definitions of stress/strain parameters result in

$$\upsilon_f = \frac{\sqrt{2}}{3}\eta_f; \quad \upsilon_c = \frac{\sqrt{2}}{3}\eta_c; \quad B = \sqrt{\frac{2}{3}}A \tag{4.40}$$

The formulation is valid as long as $f = 0$, i.e. the stress point remains on the bounding surface.

The stress reversal histories are described by invoking a two-surface formulation that bears some formal similarities to the framework described in the preceding section. In particular, the deformation process is described in terms of evolution of the *yield surface* that undergoes a kinematic hardening within the domain enclosed by the bounding surface.

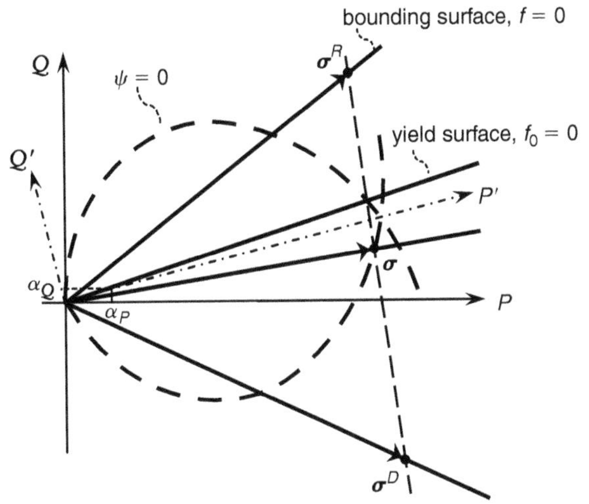

Figure 4.5 Deviatoric hardening framework; bounding and yield surfaces in 'triaxial' {P,Q} space

The equation of the yield surface is assumed in the form

$$f_0 = |Q'| - \eta_0 P' = 0 \tag{4.41}$$

Here, $\{P', Q'\}$ is a set of local axes obtained by rotation about the stress space origin. The current position of the yield surface is defined in terms of $\alpha = \{\alpha_P, \alpha_Q\}^T$, i.e. a unit vector along P'-axis, Figure 4.5. The standard transformation rule applies here, as defined in eq.(4.34).

The parameter η_0 specifies the size of the elastic domain. Note that in cohesionless granular materials that are in a loose to medium dense state of compaction, the irreversible deformations develop almost concurrently with the application of the external load. It means that, for this class of materials $\eta_0 << v_f$ and the description typically invokes the notion of evanescent elastic domain (cf. [32], [38]). Thus, in the formulation outlined here, η_0 assumes a constant threshold value. In view of this assumption, the yield surface (4.41) is taken as symmetric with respect to P'-axis. The latter is a simplification; it is consistent though with an assertion made earlier which stipulates that for $\eta \to 0$ there is $g \to 1$.

The description of plastic deformation process requires again the specification of the field of hardening moduli, together with the kinematics of the yield surface. In the present formulation, the flow rule is non-associated and it is written in the functional form analogous to representation (4.5), i.e.

$$d\varepsilon_V^p = \frac{1}{K_p}(n_P dP + n_Q dQ)\bar{n}_P; \quad d\varepsilon_Q^p = \frac{1}{K_p}(n_P dP + n_Q dQ)\bar{n}_Q \tag{4.42}$$

Here, $\boldsymbol{n}=\{n_P,n_Q\}^T$ and $\bar{\boldsymbol{n}}=\{\bar{n}_P,\bar{n}_Q\}^T$ are unit vectors normal to the yield $f_0=0$ and the local plastic potential $\psi_0=const.$ surfaces, respectively, and are defined according to eq.(4.6). The hardening modulus K_p is described by an interpolation rule (4.7)

$$K_p = K_R + K_0\left(\delta/\delta_0\right)^{\gamma};$$
$$K_R = H_p\left[\left(\frac{\partial f_0}{\partial P}\right)^2+\left(\frac{\partial f_0}{\partial Q}\right)^2\right]^{-1/2}\left[\left(\frac{\partial \psi_0}{\partial P}\right)^2+\left(\frac{\partial \psi_0}{\partial Q}\right)^2\right]^{-1/2} \tag{4.43}$$

Note that, as an alternative, a simplified form of (4.43) may be employed, as suggested in ref. [38], viz.

$$K_p = \left[K_R^{-1}\left(1-\delta/\delta_0\right)^{\gamma}\right]^{-1} \tag{4.44}$$

The latter incorporates only one material parameter and satisfies the constraint $\delta\to 0 \Rightarrow K_p \to K_R$, while for $\delta\to\delta_0$ there is $K_p\to\infty$.

The distance between the yield and the bounding surface, δ, is measured here in terms of the spatial angle between the current and the conjugate stress points. Note that within the $\{P,Q\}$ representation, a smooth transition from states on $f_0=0$ to $f=0$ is automatically ensured, so that the definition of the conjugate point $\boldsymbol{\sigma}^R=\{P^R,Q^R\}^T$ is somewhat arbitrary. Here, $\boldsymbol{\sigma}^R$ is defined as a point located at the intersection of the bounding surface and a line that passes through the current stress point and is normal to the axis of the yield surface, Figure 4.5. The coordinates of this point are

$$P^R = P+\Lambda(P-P'\alpha_P);\quad Q^R = Q+\Lambda(Q-P'\alpha_P) \tag{4.45}$$

which upon the substitution in the expression for the bounding surface, eq.(4.37), yields

$$\Lambda = [P'(\alpha_P-\alpha_Q)/f-1]^{-1} \tag{4.46}$$

Here, $f<0$ is the value of the function f, eq.(4.37), for the current stress state $\{P,Q\}$. Given the location of the conjugate point, the angle δ is now defined as

$$\delta = \cos^{-1}\frac{(\boldsymbol{\sigma})^T\boldsymbol{\sigma}^R}{|\boldsymbol{\sigma}||\boldsymbol{\sigma}^R|} = \cos^{-1}\left(\frac{PP^R+QQ^R}{(P^2+Q^2)((P^R)^2+(Q^R)^2)}\right) \tag{4.47}$$

Note that in eq.(4.37), f is defined by two distinct values of g, so that two roots for Λ are obtained from eq.(4.46). The second value provides the location of the co-called *datum* stress point $\boldsymbol{\sigma}^D$, Figure 4.5, which serves to identify the maximum spatial angle δ_0 using a similar representation to that of (4.47).

In 'triaxial' $\{P,Q\}$-space, the kinematics of the yield surface involves a simple rotation about the stress space origin. Thus,

$$\alpha_P\, d\alpha_P + \alpha_Q\, d\alpha_Q = 0 \tag{4.48}$$

Introducing this constraint in the consistency condition

$$df_0 = \frac{\partial f_0}{\partial P}dP + \frac{\partial f_0}{\partial Q}dQ + \frac{\partial f_0}{\partial \alpha_P}d\alpha_P + \frac{\partial f_0}{\partial \alpha_Q}d\alpha_Q = 0 \tag{4.49}$$

yields

$$d\alpha_P = \left(\frac{\partial f_0}{\partial P}dP + \frac{\partial f_0}{\partial Q}dQ\right) / \left(\frac{\partial f_0}{\partial \alpha_Q}\frac{\alpha_P}{\alpha_Q} - \frac{\partial f_0}{\partial \alpha_P}\right); \quad d\alpha_Q = -\frac{\alpha_P}{\alpha_Q}d\alpha_P \tag{4.50}$$

The final point that needs to be addressed here is the specification of the direction of plastic flow, viz. $\bar{\boldsymbol{n}}$ in eq.(4.42). Referring again to Figure 4.5, the local plastic potential ψ_0 is defined by engaging the current branch of the global potential surface, eq.(4.39), and its mirror image about P'-axis. Thus, incorporating the inverse transformation

$$P = P'\alpha_P - Q'\alpha_Q; \quad Q = Q'\alpha_P + P'\alpha_Q \tag{4.51}$$

and substituting in eq.(4.39), gives a parametric form

$$\psi_0 = |Q' + P'\xi| + g\upsilon_c(P' - Q'\xi)\ln\left(\frac{P' - Q'\xi}{\overline{P}}\right) = 0; \quad \xi = \alpha_Q/\alpha_P \tag{4.52}$$

where $\overline{P}$ is evaluated from the condition of $\psi_0 = 0$. Note that while $\psi_0 = 0$ is affected by the sign of Q (viz. the variable g), it remains symmetric with respect to the axis of the yield surface, which is consistent with the functional form of $f_0 = 0$, eq.(4.41).

Concluding, the constitutive relation assumes the form that is similar to representation (4.16), i.e,

$$\begin{Bmatrix} d\varepsilon_V \\ d\varepsilon_Q \end{Bmatrix} = \begin{bmatrix} \left(\frac{1}{3K} + \frac{1}{K_p}n_P\bar{n}_P\right) & \left(\frac{1}{K_p}n_Q\bar{n}_P\right) \\ \left(\frac{1}{K_p}n_P\bar{n}_Q\right) & \left(\frac{1}{2G} + \frac{1}{K_p}n_Q\bar{n}_Q\right) \end{bmatrix} \begin{Bmatrix} dP \\ dQ \end{Bmatrix} \tag{4.53}$$

subject to interpolation rule (4.43) or (4.44), and the kinematic rule (4.50).

4.2.2 Comments on the performance

In this section, some illustrative examples are given in order to assess the performance of the extended deviatoric hardening framework. In particular, a number of undrained loading programs carried out on typical cohesive granular media is considered and the focus is on modeling of liquefaction and cyclic mobility effects. The simulations incorporate the values of material parameters similar to those in Section 3.3.2, which are representative of sand in a medium dense and a dense state of compaction. Thus, for a loose/medium dense material

$$G = 35\,\text{MPa}; \quad K = 60\,\text{MPa (at } p = 500\,\text{kPa)}; \quad \eta_f = 1.2; \quad \eta_c = 1.15; \quad A = 0.001$$

while for a dense one

$$G = 45\,\text{MPa};\quad K = 100\,\text{MPa (at } p = 500\,\text{kPa)};\quad \eta_f = 1.4;\quad \eta_c = 1.2;\quad A = 0.0005$$

where K is assumed to vary linearly with p. The tests examined here involve again undrained axial compression at the confinement of 500 kPa. Thus, the conditions are the same as those depicted in Figure 3.16 and the emphasis now is on cyclic response under both stress and strain-controlled regimes. For the ease of comparison, the results are presented in terms of standard stress/strain measures, i.e. $\{p, q\}, \{\varepsilon_v, \varepsilon_q\}$.

Implementation of the anisotropic hardening rule requires information on the parameter η_0, eq.(4.41), as well as the constants appearing in the interpolation function. While the choice of η_0 is somewhat arbitrary and will not affect the results, as long as $\eta_0 << \eta$, the parameter(s) in the interpolation function (4.43) or (4.44) are normally defined by a trial and error procedure by fitting any stress reversal history. Here, η_0 was taken as $\eta_0 = 0.02$, and some typical values of $\gamma = 3$ (medium dense sand; interpolation rule (4.44)) and $K_0 = 10 \times 10^6$ kPa, $\gamma = 1$ (dense sand; interpolation rule (4.43)) were chosen.

The results of numerical simulations for loose/medium dense sand are shown in Figures 4.6 and 4.7. Figure 4.6 presents the results of a two-way strain-controlled undrained test with the axial strain amplitude of $\varepsilon_1 = \pm 0.15\%$. Both the effective stress trajectories as well as the deviatoric characteristics are given. The simulations indicate that during the test, the pore pressure progressively builds up causing the effective stress path to migrate towards the origin. The pore pressures are generated successively in both extension and compression domains and the gradual decrease of stress amplitudes is recorded. After a number of cycles the effective pressure reduces to a small residual value.

Figure 4.7 shows the response under two-way stress-controlled regime. Here, the stress amplitude is fixed at ∓ 80 kPa in extension and compression domains, respectively. The response is, in general, consistent with that shown in Figure 4.6.

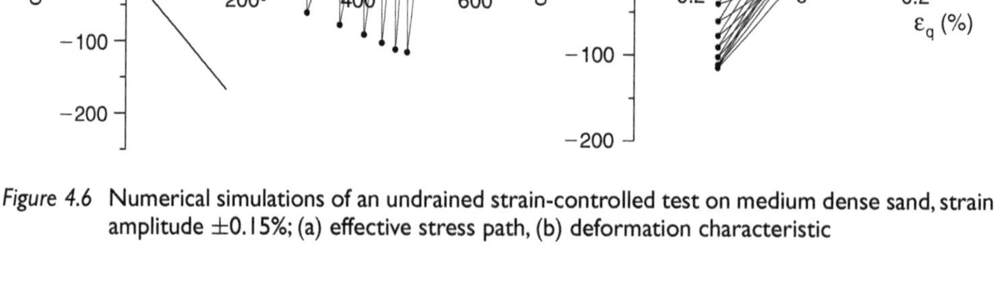

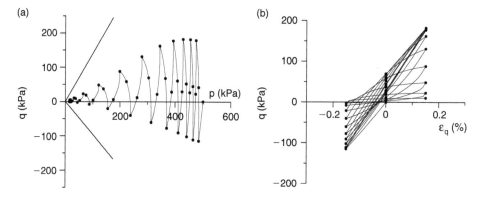

Figure 4.6 Numerical simulations of an undrained strain-controlled test on medium dense sand, strain amplitude ±0.15%; (a) effective stress path, (b) deformation characteristic

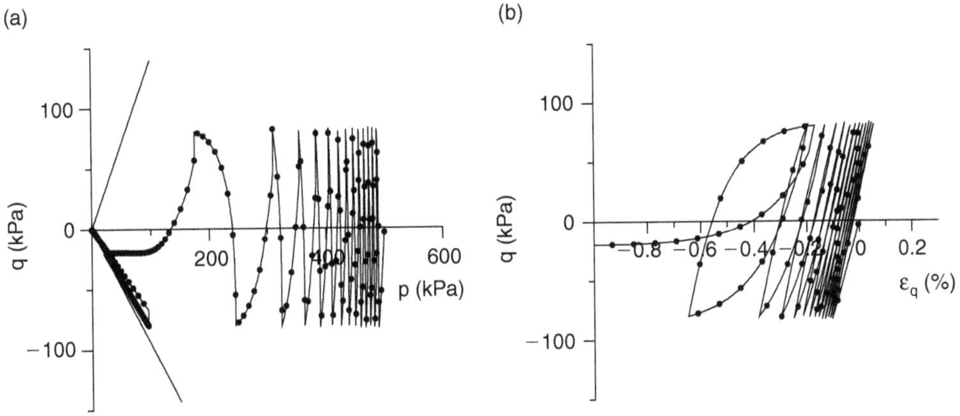

Figure 4.7 Numerical simulations of an undrained stress-controlled test on medium dense sand, stress amplitude ±80 kPa; (a) effective stress path, (b) deformation characteristic

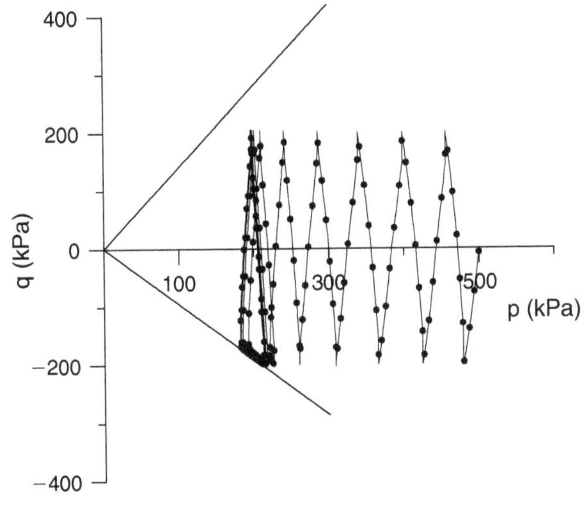

Figure 4.8 Effective stress trajectory in an undrained stress-controlled test on dense sand (20 stress-reversal cycles with amplitude ±200 kPa).

During the consecutive cycles, the effective stress trajectory migrates towards the origin and the resulting strain amplitude progressively increases. After a number of cycles, the effective pressure drops to virtually zero and the corresponding strain becomes very large. At this point, the contact between the grains is lost and the sample liquefies.

Note that the only additional parameter employed in the simulations shown in Figures 4.6 and 4.7 is γ, viz. eq.(4.44). The value of γ will, in general, affect the number of cycles to produce liquefaction; it will not impact the qualitative trends though.

Finally, Figure 4.8 presents the numerical predictions for a two-way stress-controlled test on dense sand. Here, after a number of cycles accompanied by a progressive generation of pore pressure, the effective stress trajectory rapidly stabilizes tracing a closed loop in the effective stress space. The material never liquefies. Note that the response is, in fact, similar to that depicted in Figure 4.4 using the volumetric hardening framework.

The general trends as identified here are typical of cyclic response of saturated sand at various degrees of compaction, as pointed out in Section 3.1.3. For a more comprehensive discussion on the experimental behaviour of geomaterials, the reader is again referred to Chapter 8 of this book.

4.2.3 Generalization and specification of the constitutive matrix

In order to establish the general formulation of the problem, introduce first a set of stress measures $I', \bar{\sigma}'$ associated with the coordinate system for which the stress space diagonal is along the current axis of the yield surface $f_0 = 0$. Recalling the representation introduced earlier in Chapter 2, eqs.(2.5) and (2.6), the components of the stress deviator s'_{ij} can be defined in the octahedral plane of $f_0 = 0$. Since α_{ij} is a unit vector along the axis of $f_0 = 0$, the following relations apply

$$s'_{ij} = \sigma_{ij} - I'\alpha_{ij}; \quad I' = \alpha_{ij}\sigma_{ij}; \quad \bar{\sigma}' = \left(\frac{1}{2}s'_{ij}s'_{ij}\right)^{1/2} \tag{4.54}$$

Note that if the axis of the yield surface is directed along the global stress space diagonal, the above equations reduce to

$$\alpha_{ij} = -\frac{1}{\sqrt{3}}\delta_{ij} \quad \Rightarrow \quad s'_{ij} = s_{ij}; \quad I' = -\frac{1}{\sqrt{3}}\sigma_{ii} = \sqrt{3}\sigma_m; \quad \bar{\sigma}' = \bar{\sigma} \tag{4.55}$$

i.e. define the conventional invariants. For the 'triaxial' conditions, i.e. $\sigma_2 = \sigma_3; \alpha_2 = \alpha_3$, representation (4.54) becomes

$$I' = \sigma_1\alpha_1 + 2\sigma_3\alpha_3 = P'; \quad \bar{\sigma}' = \sigma_1\alpha_3 - \sigma_3\alpha_1 = Q'/\sqrt{2} \tag{4.56}$$

Consider first the states on the bounding surface $f = 0$, eq.(4.37). Employing representation (3.44) of Section 3, which for the present choice of stress measures gives

$$P = \sqrt{3}\sigma_m; \quad Q = \sqrt{2}\bar{\sigma}/g(\theta) \tag{4.57}$$

and substituting in expressions (4.37) and (4.39), respectively, yields

$$f = \sqrt{2}\bar{\sigma} - \sqrt{3}\eta\sigma_m\, g(\theta) = 0; \quad \eta = \eta(\varepsilon^p_Q)$$
$$\psi = \sqrt{2}\bar{\sigma} + \sqrt{3}\upsilon_c\, g(\theta)\sigma_m \ln\left(\frac{\sigma_m}{\sigma^0_m}\right) = 0 \tag{4.58}$$

The parameter η depends on the history of plastic distortion ε^p_Q, which is defined by the general expression

$$\varepsilon^p_Q = \int d\varepsilon^p_Q; \quad d\varepsilon^p_Q = \pm\sqrt{J_{2\dot{\varepsilon}}} \tag{4.59}$$

Here, $J_{2\dot{\varepsilon}}$ is again the second invariant of the deviatoric part of plastic strain increment $d\boldsymbol{\varepsilon}^p$, and $d\varepsilon^p_Q > 0$ for $s_{ij}ds_{ij} > 0$. Note that, under the constraints (4.40), the above representation is identical to (3.67) and (3.68) in Section 3.3.3. Thus, the material response is governed here by the constitutive relation (3.72).

For stress reversal histories, the functional form of the yield surface $f_0 = 0$, eq.(4.41), is obtained by invoking the definitions (4.54) and (4.56). Thus,

$$f_0 = \sqrt{2}\bar{\sigma}' - \eta_0\, I' = 0 \tag{4.60}$$

In a similar way, the local plastic potential function ψ_0, eq.(4.52), can be defined as

$$\psi_0 = \sqrt{2}\bar{\sigma}' + I'\xi + g(\theta)\upsilon_c(I' - \sqrt{2}\bar{\sigma}\xi)\ln\left(\frac{I' - \sqrt{2}\bar{\sigma}\xi}{\bar{P}}\right) = 0;$$
$$\xi = -\sqrt{6J_{2\alpha}}/\mathrm{tr}(\boldsymbol{\alpha}) \tag{4.61}$$

where $J_{2\alpha}$ is the second invariant of the deviatoric part of α_{ij} and $\mathrm{tr}(\boldsymbol{\alpha}) = \alpha_{ii}$.

The flow rule (4.42) can be expressed in a general form

$$d\boldsymbol{\varepsilon}^p = \frac{1}{K_p}(\boldsymbol{n}^T d\boldsymbol{\sigma})\bar{\boldsymbol{n}}; \quad \boldsymbol{n} = \frac{\partial f_0}{\partial \boldsymbol{\sigma}} \Big/ \left|\frac{\partial f_0}{\partial \boldsymbol{\sigma}}\right|; \quad \bar{\boldsymbol{n}} = \frac{\partial \psi_0}{\partial \boldsymbol{\sigma}} \Big/ \left|\frac{\partial \psi_0}{\partial \boldsymbol{\sigma}}\right| \tag{4.62}$$

Here, K_p is defined by the interpolation function (4.42) or (4.43), in which

$$K_p = K_R + K_0\left(\delta/\delta_0\right)^{\gamma}; \quad K_R = H_p \Big/ \left(\left|\frac{\partial f_0}{\partial \boldsymbol{\sigma}}\right|\left|\frac{\partial \psi_0}{\partial \boldsymbol{\sigma}}\right|\right); \quad \delta = \cos^{-1}\frac{(\boldsymbol{\sigma})^T\boldsymbol{\sigma}^R}{|\boldsymbol{\sigma}||\boldsymbol{\sigma}^R|} \tag{4.63}$$

and H_p, i.e. the hardening modulus, is evaluated at the conjugate point $\boldsymbol{\sigma}^R$ on the bounding surface. The location of this point is obtained from a generalized form of relation (4.45). Referring to Figure 4.9a, which presents the octahedral π'-plane section (with unit normal $\boldsymbol{\alpha}$) of both the yield and bounding surface, if $\boldsymbol{\sigma}$ is the current stress state on $f_0 = 0$ then its conjugate is defined by

$$\boldsymbol{\sigma}^R = \boldsymbol{\sigma} + \Lambda(\boldsymbol{\sigma} - I'\boldsymbol{\alpha}) \tag{4.64}$$

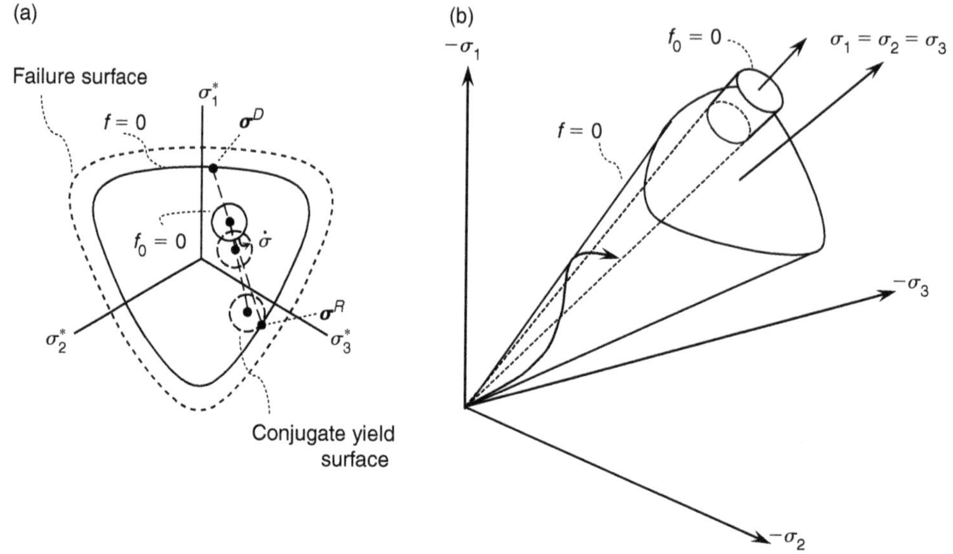

Figure 4.9 Bounding and yield surfaces in (a) octahedral plane; (b) principal stress space

which upon substitution in the equation of the bounding surface $f(\boldsymbol{\sigma}^R, \varepsilon^p_Q) = 0$, eq.(4.58), gives the roots for Λ.

The translation rule governing the kinematics of the yield surface should be formulated in such a way as to avoid the intersection with the bounding surface. For this purpose, it is convenient to envisage the so-called *conjugate yield surface*, Figure 4.9, which is tangential to the bounding surface at the conjugate point $\boldsymbol{\sigma}^R$ and its location is defined by a unit vector $\boldsymbol{\alpha}^R$

$$\boldsymbol{\alpha}^R = \left(\boldsymbol{\sigma}^R - \Lambda\frac{\partial f}{\partial \boldsymbol{\sigma}}\right) \Big/ \left|\boldsymbol{\sigma}^R - \Lambda\frac{\partial f}{\partial \boldsymbol{\sigma}}\right| \tag{4.65}$$

satisfying the equation $f_0(\boldsymbol{\sigma}^R, \boldsymbol{\alpha}^R) = 0$, eq.(4.60). The translation rule now is defined by postulating that the current yield surface moves in the plane formed by $\boldsymbol{\alpha}$ and $\boldsymbol{\alpha}^R$. Since $d\boldsymbol{\alpha}$ must be orthogonal to $\boldsymbol{\alpha}$, as the latter is a unit vector, the following expression is obtained after simple algebra

$$d\boldsymbol{\alpha} = \boldsymbol{\beta}\, d\mu; \quad \boldsymbol{\beta} = [\boldsymbol{\alpha}^R - ((\boldsymbol{\alpha})^T \boldsymbol{\alpha}^R)\boldsymbol{\alpha}] \tag{4.66}$$

where $d\mu$ is a constants which can be determined from the consistency condition $df_0 = 0$, i.e.

$$df_0 = \left(\frac{\partial f_0}{\partial \boldsymbol{\sigma}}\right)^T d\boldsymbol{\sigma} + \left(\frac{\partial f_0}{\partial \boldsymbol{\alpha}}\right)^T d\boldsymbol{\alpha} = 0 \Rightarrow \; d\mu = \left(\left(\frac{\partial f_0}{\partial \boldsymbol{\sigma}}\right)^T d\boldsymbol{\sigma}\right) \Big/ \left(\left(\frac{\partial f_0}{\partial \boldsymbol{\alpha}}\right)^T \boldsymbol{\beta}\right) \tag{4.67}$$

Finally, the constitutive relation is obtained by following the procedure analogous to that of Section 4.1.2, i.e. eqs.(4.27)–(4.29). For a strain-controlled history, the plastic strain rates (4.62) are defined as

$$d\boldsymbol{\varepsilon}^p = \frac{1}{\tilde{H}}(\boldsymbol{n}^T[D^e]\,d\boldsymbol{\varepsilon})\bar{\boldsymbol{n}}; \quad \tilde{H} = (H_e + H_p)\Big/\left(\left|\frac{\partial f_0}{\partial \boldsymbol{\sigma}}\right|\left|\frac{\partial \psi_0}{\partial \boldsymbol{\sigma}}\right|\right) \tag{4.68}$$

Thus, invoking the elastic constitutive relation, together with the additivity postulate, yields

$$d\boldsymbol{\sigma} = [D^e](d\boldsymbol{\varepsilon} - d\boldsymbol{\varepsilon}^p) = [D^{ep}]d\boldsymbol{\varepsilon}; \quad [D^{ep}] = [D^e] - \frac{1}{\tilde{H}}([D^e]\,\bar{\boldsymbol{n}}\,\boldsymbol{n}^T[D^e]) \tag{4.69}$$

It is worth emphasizing, once again, that the framework outlined here, although numerically complex, requires a limited number of material parameters. In fact, only one additional constant γ is needed when the interpolation function (4.44) is employed. This is certainly an advantage, given that the predictive abilities of this framework, as examined in Section 4.2.2, appear to be quite reasonable.

Chapter 5

Numerical integration of constitutive relations

In previous chapters, various plasticity frameworks have been outlined for modelling of the mechanical response of geomaterials. Those included a class of elastic-perfectly plastic formulations as well as the approaches incorporating isotropic strain-hardening and combined isotropic-kinematic hardening rules. Unlike the linear elasticity, the constitutive relations in elastoplasticity are not analytically integrable; i.e. they cannot be reduced to finite relations between stress and strain tensors. Thus, for an arbitrary stress/strain history, a numerical integration is required in order to define the resulting mechanical response. In this chapter, basic techniques for numerical integration of elastoplastic constitutive equations are briefly reviewed. Note that only point integration algorithms are dealt with here; the numerical techniques for analyzing the boundary-value problems are discussed elsewhere (see, e.g., Simo and Hughes [39]).

5.1 EULER'S INTEGRATION SCHEMES

The numerical integration of incremental plasticity relations represents the following nonlinear problem:

– Starting from a known state of stress $\boldsymbol{\sigma}$, strain $\boldsymbol{\varepsilon}$ and internal plastic variables $\boldsymbol{\alpha}$, i.e. $\{\boldsymbol{\sigma}, \boldsymbol{\varepsilon}, \boldsymbol{\alpha}\}_n$ at step n, find their respective increments, for a prescribed loading history, that are consistent with the constitutive law. Given those, update the current values to $\{\boldsymbol{\sigma}, \boldsymbol{\varepsilon}, \boldsymbol{\alpha}\}_{n+1}$ at step $n+1$.

Before formulating general procedures for dealing with arbitrary paths in the stress/strain space, let us first recall the basic schemes for numerical integration. The simplest form of numerical integration is the Euler's explicit method. It involves evaluation of the derivative of a function at a certain step and linear extrapolation, based on that derivative, to the next step. Alternative approaches involve midpoint and implicit (backward) schemes. In order to illustrate these procedures, consider the evolution of a variable $y(t)$ during a time interval $t \in [0, T]$. Let the initial-value problem be given by the differential equation

$$\dot{y} = f(y) \tag{5.1}$$

subject to $y(t=0) = y_o$.

For the purpose of numerical integration, the time interval is divided into N subintervals. Let the value of y at the step t_n be known. The basic notion of Euler's scheme is to assume a linear variation of the variable y during the time step $[t_n, t_n + \Delta t]$. Thus, the time derivative is approximated by difference Δy evaluated over Δt, i.e. $\dot{y} \approx \Delta y / \Delta t$. The incremental value Δy can be expressed in the following form:

$$\Delta y = \Delta t\{(1-\chi)f_n + \chi f_{n+1}\}, \chi \in [0,1] \tag{5.2}$$

By assigning now a particular value to the coefficient χ, various difference schemes may be obtained. In practice, the following three schemes are commonly used:

$$\chi = 0 - \text{explicit (forward)};\quad \chi = 0.5 - \text{implicit (midpoint)};$$
$$\chi = 1 - \text{implicit (backward)}$$

Note that for the explicit scheme, the incremental value Δy depends only on the known value of the function f_n at the previous step $t = t_n$. The discrete equation (5.2) can then be explicitly integrated. For implicit schemes, however, the value of f_{n+1} at the current step $t_n + \Delta t$ is unknown so that an iterative procedure is required.

In what follows, different techniques for integration of plasticity-based relations are reviewed that incorporate Euler's schemes. It should be emphasized again that the standard plasticity framework is *time-independent*, so that the integration is, in fact, carried out along a prescribed stress/strain trajectory. For the sake of clarity, the presentation is focused first on the integration in $\{p,q\}$-space. Some examples are given to assess the accuracy of different schemes in the context of volumetric and deviatoric hardening frameworks. The chapter is concluded by discussing general procedures for integration along arbitrary strain trajectories.

5.2 NUMERICAL INTEGRATION OF $\{p,q\}$ FORMULATION

Let us examine the basic point integration schemes in the context of the framework of deviatoric hardening, as outlined in Section 3.3 of Chapter 3. To recall, the key expressions defining the active loading (f) and plastic potential (ψ) functions are

$$\begin{cases} f = q - \eta p = 0; \quad \eta = \eta_f \dfrac{\varepsilon_q^p}{A + \varepsilon_q^p} \\ \psi = q + \eta_c p \ln\left(\dfrac{p}{\bar{p}}\right) = 0 \end{cases} \tag{5.3}$$

Assume that the primary variables $\{p, q, \varepsilon_v, \varepsilon_q, \varepsilon_v^p, \varepsilon_q^p\}$ are known at the step n. The problem then is to evaluate the incremental variations of these variables along a prescribed loading path. In general, the attention should be given to stress/strain trajectories that are typical of both drained and undrained tests, i.e. involve either stress or strain-controlled conditions, respectively. Both these cases are addressed below.

Table 5.1 Algorithm for explicit direct integration (stress-controlled problems)

Known $\{p, q, \varepsilon_v, \varepsilon_q, \varepsilon_v^p, \varepsilon_q^p\}$ at the step n, and $f(p^{(n)}, q^{(n)}, \varepsilon_q^{p(n)}) = 0$

1. Given: $\{\Delta p, \Delta q\}$
2. Update: $p^{(n+1)} = p^{(n)} + \Delta p, \quad q^{(n+1)} = q^{(n)} + \Delta q$
3. Evaluate the incremental value of the hardening function:

$$f = q^{(n+1)} - (\eta^{(n)} + \Delta\eta)\, p^{(n+1)} = 0, \quad \Delta\eta = \frac{q^{(n+1)} - \eta^{(n)} p^{(n+1)}}{p^{(n+1)}}$$

4. Compute:

$$\begin{cases} \Delta\varepsilon_q^p = \dfrac{\Delta\eta}{\partial\eta/\partial\varepsilon_q^{p(n)}}, & \Delta\varepsilon_v^p = \Delta\varepsilon_q^p \dfrac{\partial\psi/\partial p^{(n)}}{\partial\psi/\partial q^{(n)}} \\ \Delta\varepsilon_v^e = \dfrac{\Delta p}{K}, & \Delta\varepsilon_q^e = \dfrac{\Delta q}{3G} \end{cases}$$

5. Update:

$$\varepsilon_v^{p(n+1)} = \varepsilon_v^{p(n)} + \Delta\varepsilon_v^p, \quad \varepsilon_q^{p(n+1)} = \varepsilon_q^{p(n)} + \Delta\varepsilon_q^p$$

$$\varepsilon_v^{(n+1)} = \varepsilon_v^{(n)} + \Delta\varepsilon_v^e + \Delta\varepsilon_v^p, \quad \varepsilon_q^{(n+1)} = \varepsilon_q^{(n)} + \Delta\varepsilon_q^e + \Delta\varepsilon_q^p$$

6. Set $n = n + 1$; go to (1) for the next step

5.2.1 Stress-controlled scheme

Assume that the stress trajectory is specified so that the task is to define the corresponding response in strain. Denote the current stress/strain state as $\boldsymbol{\sigma} = \{p, q\}^T$, $\boldsymbol{\varepsilon} = \{\varepsilon_v, \varepsilon_q\}^T$. In this case,

$$\boldsymbol{\varepsilon} = \int_t \dot{\boldsymbol{\varepsilon}}\, dt = \int_{\boldsymbol{\varepsilon}} d\boldsymbol{\varepsilon} = \int_{\boldsymbol{\varepsilon}} (d\boldsymbol{\varepsilon}^e + d\boldsymbol{\varepsilon}^p) = [\mathbf{C}^e]\boldsymbol{\sigma} + \frac{1}{H_p}\int_{\boldsymbol{\sigma}} \left(\frac{\partial\psi}{\partial\boldsymbol{\sigma}}\right)^T \frac{\partial f}{\partial\boldsymbol{\sigma}}\, d\boldsymbol{\sigma} \tag{5.4}$$

where $H_p > 0$ is the plastic hardening modulus defined in eq.(3.66).

It is evident from (5.4) that time t is a fictitious variable that is not explicitly involved in the formulation. Thus, the integration in (5.4) is performed along a prescribed stress trajectory which is incrementally divided into a number of steps. Note that, since the values of $\{p,q\}$ are known at each loading step n, it is efficient to employ a direct integration using the explicit Euler's scheme. The flow chart for the algorithm is given in Table 5.1. The procedure outlined in this table corresponds to an active loading process, i.e. at the step n, there is $f(p^{(n)}, q^{(n)}, \varepsilon_q^{p(n)}) = 0$ and $f(p^{(n)} + \Delta p, q^{(n)} + \Delta q, \varepsilon_q^{p(n)}) > 0$.

5.2.2 Strain-controlled schemes

This is the most commonly used integration scheme. For a point integration, it can represent the test conditions associated with isochoric deformation ($\varepsilon_v = 0$), e.g. undrained response of granular media. The same scheme is required for implementation of the constitutive relation in a finite element code (cf. Ref. [39]).

In a strain-controlled problem, the strain trajectory is prescribed and the objective is to define the corresponding response in stress. In this case, using the notation of eq.(5.4), i.e. $\boldsymbol{\sigma} = \{p, q\}^T$, $\boldsymbol{\varepsilon} = \{\varepsilon_v, \varepsilon_q\}^T$, there is

$$\boldsymbol{\sigma} = \int_t \dot{\boldsymbol{\sigma}}\, dt = \int_\sigma d\boldsymbol{\sigma} = \int_\varepsilon [D^e](d\boldsymbol{\varepsilon} - d\boldsymbol{\varepsilon}^p) = [D^e]\,\boldsymbol{\varepsilon} - \int_\varepsilon [D^e]\frac{\partial \psi}{\partial \boldsymbol{\sigma}}\, d\lambda \tag{5.5}$$

where

$$d\lambda = \frac{1}{H}\left(\frac{\partial f}{\partial \boldsymbol{\sigma}}\right)^T [D^e]\, d\boldsymbol{\varepsilon}; \qquad [D^e] = \begin{bmatrix} K & 0 \\ 0 & 3G \end{bmatrix} \tag{5.6}$$

with $H = H_e + H_p$ defined in eq.(3.57).

Once again, the time does not explicitly enter the formulation so that the integration is carried out along a prescribed strain trajectory. In the integration process, the incremental values of strain $\{\Delta\varepsilon_v, \Delta\varepsilon_q\}$ are given for each load step. The response in stress is found by evaluating the plastic multiplier $\Delta\lambda$ and then plastic strain increments $\{\Delta\varepsilon_v^p, \Delta\varepsilon_q^p\}$. This, in turn, allows to determine the corresponding incremental values of stress $\{\Delta p, \Delta q\}$. In what follows, two different techniques are briefly described, viz. direct integration and the return mapping algorithm.

(i) Explicit direct integration

This method is based on the explicit Euler's scheme. A step by step integration is carried out using the known values of the variables at the step n. Thus,

$$\Delta\varepsilon_v^p = \Delta\lambda \frac{\partial \psi}{\partial p^{(n)}}, \; \Delta\varepsilon_q^p = \Delta\lambda \frac{\partial \psi}{\partial q^{(n)}}, \quad \Delta\lambda = \left(\frac{\partial f}{\partial p^{(n)}} K \Delta\varepsilon_v + 3G \frac{\partial f}{\partial q^{(n)}} \Delta\varepsilon_q\right) / H^{(n)} \geq 0 \tag{5.7}$$

The algorithmic flowchart for an active loading process, i.e. $f(p^{(n)}, q^{(n)}, \varepsilon_q^{p(n)}) = 0$ and $f(p^{(n)} + K\Delta\varepsilon_v, q^{(n)} + 3G\Delta\varepsilon_q, \varepsilon_q^{p(n)}) > 0$, is given in Table 5.2.

(ii) Return mapping scheme

The return mapping algorithms are very common in the numerical integration of inelastic constitutive relations. Before discussing a general procedure suitable for arbitrary paths in the strain space, a simplified algorithm is presented here for $p - q$ plasticity framework. The algorithm involves two primary steps that can be summarized as follows

- introduce a trial stress state; this is usually identified with the elastic solution for a given strain increment
- if the trial state is outside the yield/loading surface, use a return mapping algorithm to bring this state back onto the surface.

Due to the non-linearity of constitutive equations, the projection point of the trial state onto the current loading surface can not be found in an explicit way. It is therefore

Table 5.2 Algorithm for explicit direct integration (strain-controlled problems).

Known $\{p, q, \varepsilon_v, \varepsilon_q, \varepsilon_v^p, \varepsilon_q^p\}$ at the step n, and $f(p^{(n)}, q^{(n)}, \varepsilon_q^{p(n)}) = 0$

1. Given: $\{\Delta\varepsilon_v, \Delta\varepsilon_q\}$
2. Compute the plastic multiplier:
$$\Delta\lambda = \frac{\dfrac{\partial f}{\partial p^{(n)}} K\Delta\varepsilon_v + \dfrac{\partial f}{\partial q^{(n)}} 3G\Delta\varepsilon_q}{H_e^{(n)} + H_p^{(n)}}$$
3. Compute plastic strain increments:
$$\Delta\varepsilon_v^p = \Delta\lambda \frac{\partial\psi}{\partial p^{(n)}}, \quad \Delta\varepsilon_q^p = \Delta\lambda \frac{\partial\psi}{\partial q^{(n)}}$$
4. Update the stress state:
$$\Delta p = K(\Delta\varepsilon_v - \Delta\varepsilon_v^p), \quad \Delta q = 3G(\Delta\varepsilon_q - \Delta\varepsilon_q^p)$$
$$p^{(n+1)} = p^{(n)} + \Delta p, \quad q^{(n+1)} = q^{(n)} + \Delta q$$
$$\varepsilon_q^{p(n+1)} = \varepsilon_q^{p(n)} + \Delta\varepsilon_q^p$$
$$\varepsilon_v^{(n+1)} = \varepsilon_v^{(n)} + \Delta\varepsilon_v, \quad \varepsilon_q^{(n+1)} = \varepsilon_q^{(n)} + \Delta\varepsilon_q$$
5. Update the hardening parameter and check the loading function f:
$$\eta^{(n+1)} = \eta_f \frac{\varepsilon_q^{p(n+1)}}{A + \varepsilon_q^{p(n+1)}}$$
$$f = \left(q^{(n+1)} + \beta\frac{\partial f}{\partial q^{(n+1)}}\right) - \eta^{(n+1)}\left(p^{(n+1)} + \beta\frac{\partial f}{\partial p^{(n+1)}}\right) = 0 \quad \Rightarrow \beta = \frac{\eta p - q}{1 + \eta^2}$$
6. Correct the stress state according to:
$$p^{(n+1)} = p^{(n+1)} - \beta\eta^{(n+1)}, \quad q^{(n+1)} = q^{(n+1)} + \beta$$
7. Set $n = n + 1$; go to (1) for the next step

necessary to determine it using an iterative procedure. The return mapping algorithm, in the context of the deviatoric hardening, is summarized in Table 5.3.

5.3 NUMERICAL EXAMPLES OF INTEGRATION IN $\{p,q\}$ SPACE

In this section, some numerical examples are given in order to illustrate the performance of different integration schemes. Some typical loading paths are examined, viz. drained and undrained effective stress trajectories, and the response is compared with analytical solutions. The simulations presented here are based on both the Critical State and deviatoric hardening frameworks, as discussed in Sections 3.2 and 3.3, respectively.

5.3.1 Critical state model; drained $p = const.$ compression

In this example, the mechanical response in a drained compression test is investigated. Starting from the initial confining pressure p_0, the sample is subjected to an increase in the deviatoric stress under a constant mean stress, until the conditions at failure are reached, i.e. $q = \eta_f p$.

For the stress path considered here, i.e. $p = p_0 = const.$, an analytical solution is found using the Critical State framework outlined in Section 3.2. Setting $dp = 0$ in

Table 5.3 Return mapping algorithm for $q-p$ plasticity with deviatoric hardening.

Known $\{p, q, \varepsilon_v, \varepsilon_q, \varepsilon_v^p, \varepsilon_q^p\}$ at the step n, and $f(p^{(n)}, q^{(n)}, \varepsilon_q^{p(n)}) = 0$

1. Given: $\{\Delta\varepsilon_v, \Delta\varepsilon_q\}$
2. Compute elastic trial stress state and start iterative procedure, $k = 1$:
   $p^{trial} = p^{(n)} + K\Delta\varepsilon_v, \quad q^{trial} = q^{(n)} + 3G\Delta\varepsilon_q$
   $p^{(n+1,k)} = p^{trial}, \quad q^{(n+1,k)} = q^{trial}, \quad \varepsilon_q^{p(n+1,k)} = \varepsilon_q^{p(n)}$
3. Check the loading condition and compute the plastic multiplier:
   $$\eta^{(n+1,k)} = \eta_f \frac{\varepsilon_q^{p(n+1,k)}}{A + \varepsilon_q^{p(n+1,k)}}, \quad f^{(k)} = q^{(n+1,k)} - \eta^{(n+1,k-1)}\, p^{(n+1,k)}$$
   If $f^{(k)} \le 0$ then
   set $\Delta\lambda = 0$; go to (7)
   else
   $$\Delta\lambda = \frac{f^{(k)}}{H_e^{(k)} + H_p^{(k)}} > 0$$
4. Compute plastic strain increments and the hardening variable:
   $$\Delta\varepsilon_v^p = \Delta\lambda \frac{\partial\psi}{\partial p^{(n+1,k)}}, \quad \Delta\varepsilon_q^p = \Delta\lambda \frac{\partial\psi}{\partial q^{(n+1,k)}}, \quad \varepsilon_q^{p(n+1,k+1)} = \varepsilon_q^{p(n+1,k)} + \Delta\varepsilon_q^p$$
5. Apply the plastic correction for stress:
   $p^{(n+1,k+1)} = p^{(n+1,k)} - K\Delta\varepsilon_v^p, \quad q^{(n+1,k+1)} = q^{(n+1,k)} - 3G\Delta\varepsilon_q^p$
6. Set $k = k + 1$; go to (3)
7. Update the values of variables after convergence:
   $p^{(n+1)} = p^{(n+1,k+1)}, \quad q^{(n+1)} = q^{(n+1,k+1)}, \quad \varepsilon_q^{p(n+1)} = \varepsilon_q^{p(n+1,k+1)}$
   $\varepsilon_v^{(n+1)} = \varepsilon_v^{(n)} + \Delta\varepsilon_v, \quad \varepsilon_q^{(n+1)} = \varepsilon_q^{(n)} + \Delta\varepsilon_q$
8. Set $n = n + 1$; go to (1) for the next step

eq.(3.30) and utilizing the expression (3.33), the following closed-form representation can be obtained for the volumetric (ε_v) and deviatoric strains (ε_q)

$$\varepsilon_v = \frac{1}{C}\ln\left(\frac{\eta_f^2 p^2 + q^2}{\eta_f^2 p^2}\right) \tag{5.8}$$

$$\begin{aligned}\varepsilon_q = {} & \frac{q}{3G} - \frac{2}{C\eta_f}\tan^{-1}\left(\frac{q}{\eta_f p}\right) - \frac{6+\eta_f}{6C\eta_f}\ln\left(\frac{q-\eta_f p}{-\eta_f p}\right) \\ & + \left(\frac{1}{\eta_f} - \frac{1}{6}\right)\ln\left(\frac{\eta_f p + q}{\eta_f p}\right) - \frac{1}{6C}\ln\left(\frac{\eta_f^2 p^2 + q^2}{\eta_f^2 p^2}\right) + \frac{1}{6C}\ln\left(\frac{q^4 - \eta_f^4 p^4}{-\eta_f^4 p^4}\right)\end{aligned} \tag{5.9}$$

where $C = (1 + e_0)/(\lambda - \kappa)$.

Figure 5.1 presents a comparison between the numerical and analytical solutions. The results correspond to $p_0 = 500$ kPa, while the material parameters are the same as

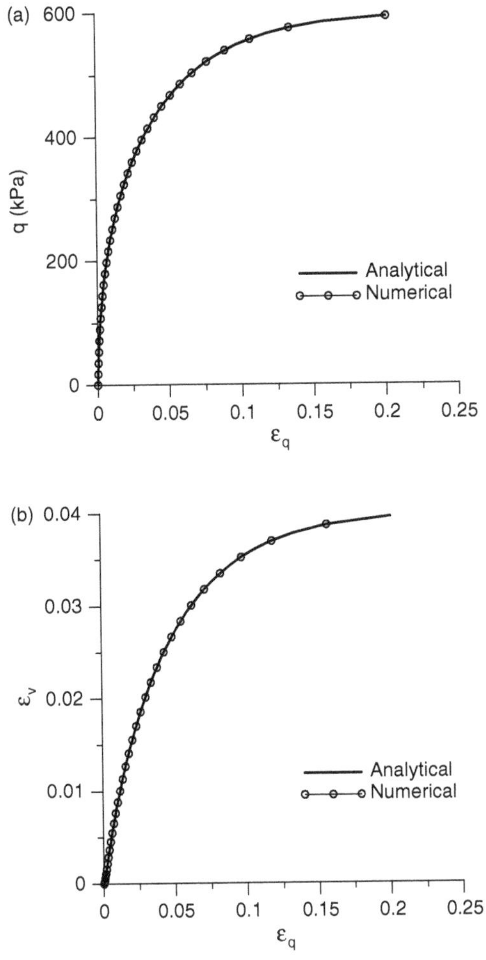

Figure 5.1 Drained $p = const.$ compression; comparison between analytical solution and numerical prediction based on return mapping algorithm (Critical State model) (a) deviatoric stress-strain characteristic, (b) volume change response

those employed earlier in Sections 3.2.2 and 4.1.2, i.e.

$$G = 30\,\text{MPa}, \quad \phi = 30^\circ, \quad \lambda = 0.13, \quad \kappa = 0.02, \quad e_0 = 0.9$$

Note that in this case, since the stress trajectory is given a priori, the explicit direct integration will always yield the results that are consistent with the analytical solution, provided the increments are sufficiently small. Therefore, the focus here is on the predictions based on the return mapping algorithm. In order to test this algorithm, an elastic solution is employed to define the strain increments, which are subsequently used in an iterative loop of the return mapping algorithm (Table 5.3). The numerical simulations shown in Figure 5.1 correspond to 100 steps. The agreement is, in general, very good. When the stress trajectory approaches the Critical State line, large deformations develop and the numerical convergence in the return mapping scheme becomes very slow, i.e. a large number of iterations is required.

5.3.2 Deviatoric hardening model; drained 'triaxial' compression

The next example involves the numerical simulation of a drained 'triaxial' compression test performed at the same initial confining pressure of $p_0 = 500\,\text{kPa}$. The simulation corresponds to the deviatoric hardening model of Section 3.3. In order to obtain a simple analytical solution, the framework is modified by assuming a linear form of the plastic potential function that results in no plastic volume change, i.e $\psi = q = const. \Rightarrow \varepsilon_v^p = 0$.

The stress trajectory is defined as $dq = 3dp$. Starting from an initial confinement of p_0, the deviatoric stress q is progressively increased, while the corresponding mean stress is evaluated as $p = p_0 + q/3$. Taking the initial confining stress as the reference state for the deformation, the volumetric and deviatoric strains can be expressed as explicit functions of stress measures p and q. Since $\varepsilon_v^p = 0$, one obtains

$$\varepsilon_v = \varepsilon_v^e = \frac{p - p_0}{K}; \quad \varepsilon_q = \frac{q}{3G} + \varepsilon_q^p \tag{5.10}$$

where

$$\varepsilon_q^p = \frac{\eta A}{\eta_f - \eta}; \quad \eta = \frac{q}{p} \tag{5.11}$$

Figure 5.2 shows the comparison between the numerical and analytical solution. The results correspond to a set of material parameters representative of a medium dense sand (Section 3.3.2), i.e.

$$G = 35\,\text{MPa}; \quad K = 60\,\text{MPa}; \quad \eta_f = 1.2; \quad A = 0.001$$

The numerical simulations are again performed using the return mapping algorithm with an iterative scheme driven by strain increments based on the elastic solution. Once again, the analytical curve is quite well represented using the same number of steps as before, i.e. 100.

5.3.3 Deviatoric hardening model; undrained 'triaxial' compression

In this section, the undrained 'triaxial' tests are simulated for specimens of very loose and medium dense sand. Again, the numerical analysis is conducted using the deviatoric hardening model. The tests are the same as those discussed in Section 3.3.2 (viz. Figures 3.16c and 3.16b), i.e. are based on the same set of material parameters and involve

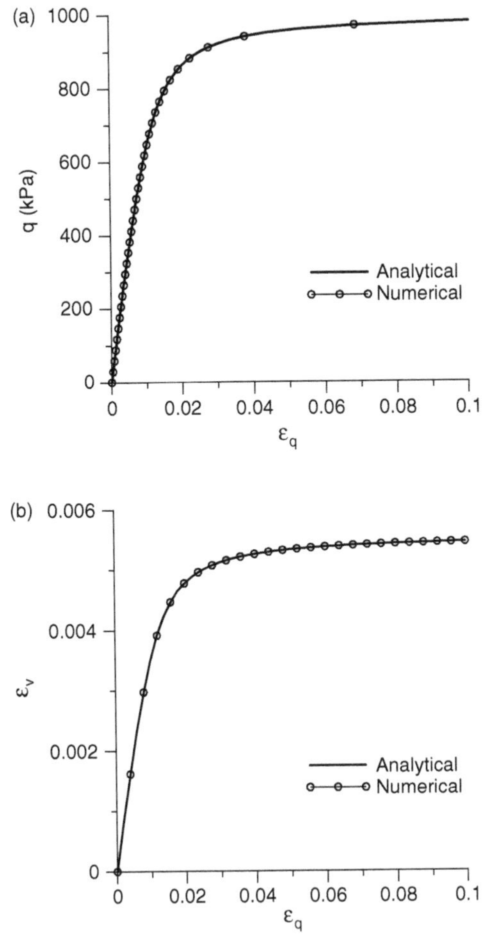

Figure 5.2 Drained 'triaxial' compression; comparison between analytical solution and numerical prediction based on return mapping algorithm (deviatoric hardening model) (a) deviatoric stress-strain characteristic, (b) volume change response

compression at the initial confining pressure of $p_0 = 500\,\text{kPa}$. The objective here is to examine the performance of different integration schemes.

The undrained conditions in granular media are enforced by imposing the kinematic constraint $\varepsilon_v = 0$. In this case, it is possible to find an analytical solution in an implicit form. Incorporating the constraint of no volume change in eq.(3.58) and utilizing representations (3.61) and (3.64), the following equation of effective stress trajectory is obtained

$$-\ln\left[\frac{\eta_f - q/p}{\eta_f}\right] - \frac{\eta_f - \eta_c}{\eta_f - q/p} + \frac{\eta_f - \eta_c}{\eta_f} = -\frac{p - p_0}{AK\eta_f} \tag{5.12}$$

At the same time, the deviatoric strain can be evaluated directly from eq.(3.62), viz.

$$\varepsilon_q = \frac{q}{3G} + \frac{(q/p)A}{\eta_f - q/p} \tag{5.13}$$

In the numerical analysis, two different integration schemes have been used, namely the explicit direct integration (Table 5.2) and the return mapping scheme (Table 5.3). Given the kinematic constraint $\varepsilon_v = 0$, the problem can be considered as strain-controlled. Figures 5.3 and 5.4 show the numerical solution for very loose sand based on 50 increments over the strain range of 5%. Figure 5.3 gives the deviatoric stress-strain characteristics, while Figure 5.4 presents the corresponding effective stress trajectories. According to the analytical solution, the deviatoric stress first increases and then progressively decreases as the stress path progresses towards the origin. The static liquefaction is produced when the effective pressure reduces to zero. It is seen from these figures that the explicit direct integration is very effecient here; i.e. 50 steps are sufficient to reproduce both analytical curves. However, the return mapping algorithm is not even able to match the qualitative trends, as the number of steps employed here is not large enough. The procedure converges to a solution that is inconsistent with the analytical one, i.e. the deviatoric stress continuously increases and no liquefaction is produced.

Figures 5.5 and 5.6 examine the influence of the number of steps on the nature of the solution based on return mapping scheme. It is evident that the accuracy improves as the number of steps increases. At the same time though, nearly 10,000 steps are needed to reproduce the analytical solution in a satisfactory manner.

Another example given here concerns a medium dense sand subjected to the same loading history. Again, the explicit direct integration and the return mapping schemes

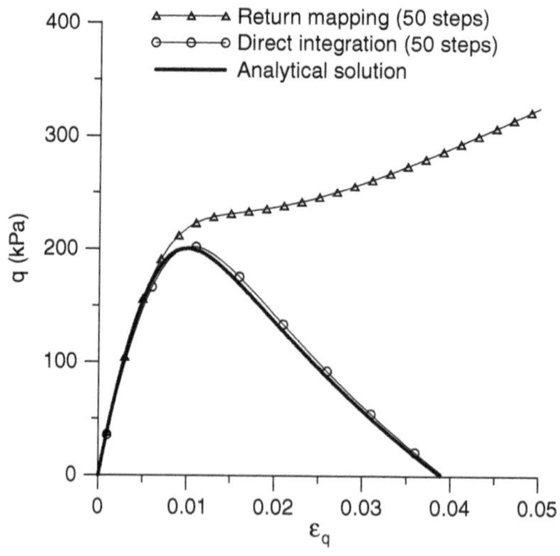

Figure 5.3 Deviatoric stress-strain characteristics in undrained compression test on a very loose sand; comparison between analytical solution and numerical predictions

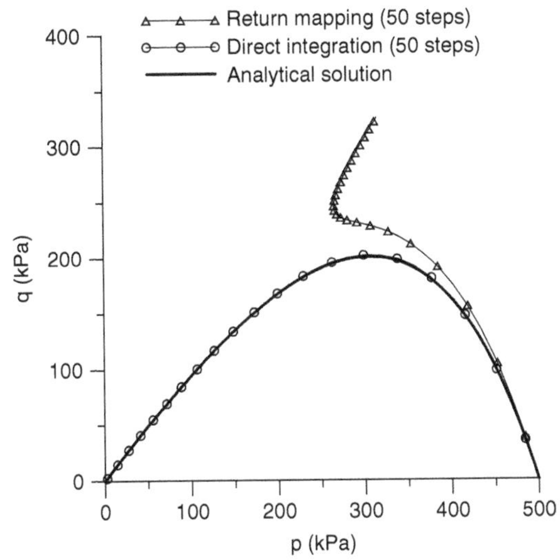

Figure 5.4 Effective stress trajectories in undrained compression test on a very loose sand; comparison between analytical solution and numerical predictions

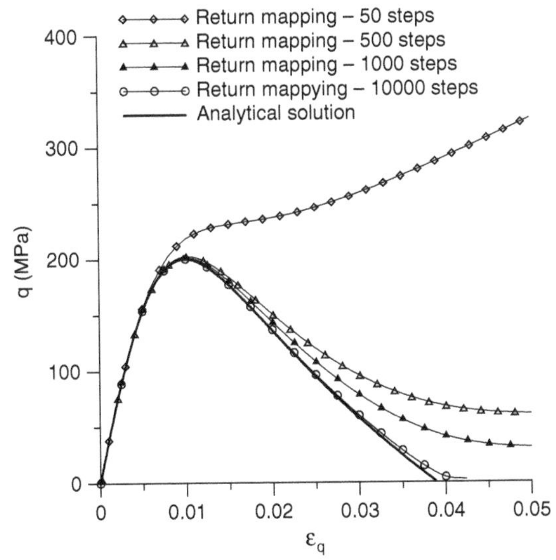

Figure 5.5 Deviatoric stress-strain characteristics in undrained compression test on a very loose sand; influence of number of steps in return mapping algorithm

have been employed to simulate the test conditions. The results, in terms of deviatoric stress-strain response and the effective stress trajectories, are shown in Figures 5.7–5.10. Again, the performance of the explicit direct integration is quite satisfactory within the range of 100–200 steps, which is evidenced in Figures 5.7 and 5.8. At the

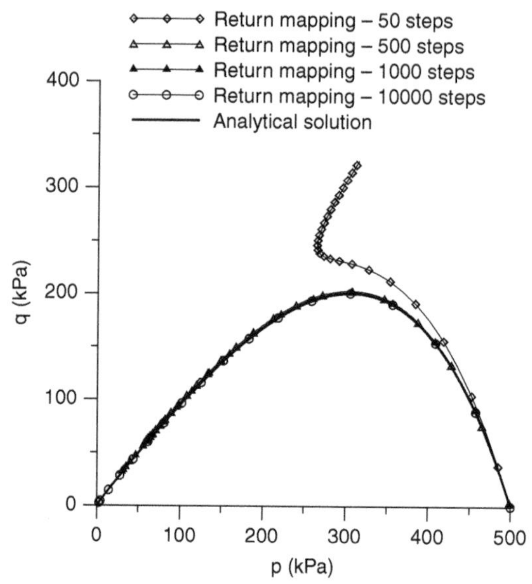

Figure 5.6 Effective stress trajectories in undrained compression test on a very loose sand; influence of number of steps in return mapping algorithm

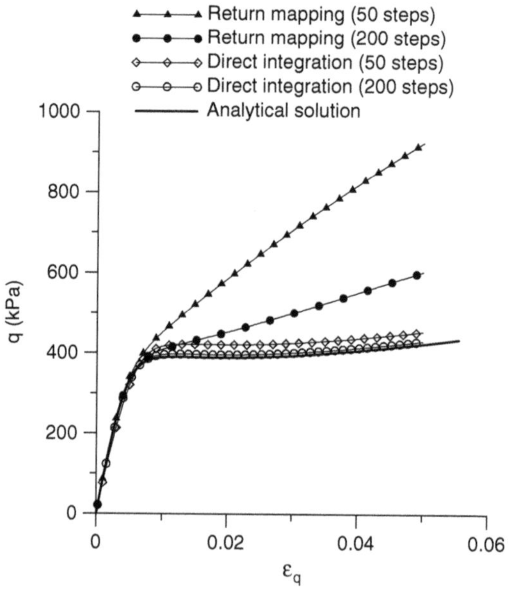

Figure 5.7 Deviatoric stress-strain characteristics in undrained compression test on a medium dense sand; comparison between analytical solution and numerical predictions

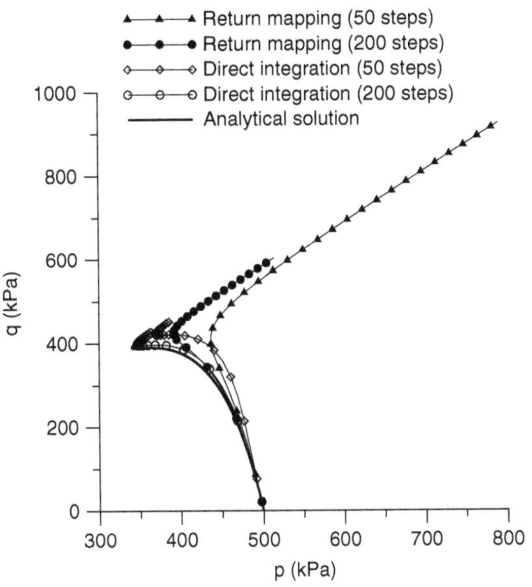

Figure 5.8 Effective stress trajectories in undrained compression test on a medium dense sand; comparison between analytical solution and numerical predictions

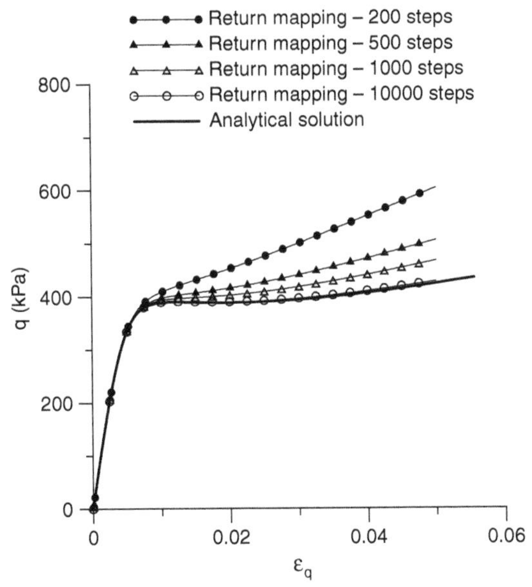

Figure 5.9 Deviatoric stress-strain characteristics in undrained compression test on a medium dense sand; influence of number of steps in return mapping algorithm

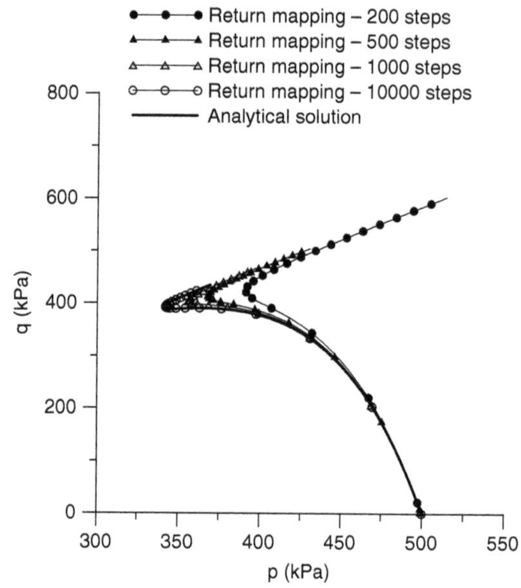

Figure 5.10 Effective stress trajectories in undrained compression test on a medium dense sand; influence of number of steps in return mapping algorithm

same time, the return mapping algorithm does not converge to the right solution within the prescribed number of increments. As the latter is progressively increased, a good accuracy is eventually achieved, as shown in Figures 5.9 and 5.10.

Given the examples presented here, it is evident that the numerical integration should always be approached with caution. The return mapping scheme is formulated in a mathematically elegant way and it is, in general, efficient and numerically stable. In certain cases, however, one needs to ensure that it converges to the solution that is mathematically and physically correct.

5.4 GENERAL METHODS FOR NUMERICAL INTEGRATION

Let us now focus on the formulation of a general procedure for numerical integration along arbitrary trajectories in the stress/strain space. Note that in most standard techniques for analysis of initial boundary-value problems, such as finite elements or finite differences, displacement rates are chosen as the basic unknowns. In this case, the local strain rates are evaluated from the kinematic relations and subsequently used in the constitutive law to determine the response in the stress rates. Therefore, the integration process in the constitutive relation itself is strain-driven. This section provides a brief review of the basic techniques of numerical integration under a strain-controlled regime.

5.4.1 Statement of algorithmic problem

For clarity of the presentation, let us centre our attention on a class of elastoplastic models with isotropic hardening. Although this is rather restrictive, the basic concepts and general procedures outlined here can be adopted to incorporate more advanced frameworks, such as for example isotropic-kinematic hardening. In formulating the numerical procedures, the elastic behaviour, which employs the operator $[D^e]$, is assumed to be linear. Furthermore, a non-associated flow rule is postulated.

Assume that the integration is to be carried out at a point $\boldsymbol{x} \in \Omega \subset R^3$ in a continuous body. Let at this point the basic variables $\{\boldsymbol{\varepsilon}_n, \boldsymbol{\varepsilon}^p_n, \boldsymbol{\alpha}_n\}$ be known at the current time step $t_n \in [0, T]$. Note that elastic strain and stress tensors are considered as dependent variables as they can be determined from

$$\boldsymbol{\varepsilon}^e_n = \boldsymbol{\varepsilon}_n - \boldsymbol{\varepsilon}^p_n; \quad \boldsymbol{\sigma} = [D^e]\left(\boldsymbol{\varepsilon}_n - \boldsymbol{\varepsilon}^p_n\right) \tag{5.14}$$

Let $\dot{\boldsymbol{\varepsilon}}$ be the given strain rate during the interval $t \in [t_n, t_n + \Delta t]$. The basic problem to be solved here is to update the variables to the step $t_{n+1} = (t_n + \Delta t) \in [0, T]$, i.e. evaluate $\{\boldsymbol{\varepsilon}_{n+1}, \boldsymbol{\varepsilon}^p_{n+1}, \boldsymbol{\alpha}_{n+1}\}$ in a manner consistent with the elastoplastic constitutive relation.

In the statement above, the problem is formally phrased in terms of time t. As mentioned earlier, although the general formulation of any initial boundary-value problem invokes time, the integration of the constitutive law is usually carried out over a prescribed strain trajectory, so that time is a fictitious variable. It should be noted, however, that this is not a general case. In fact, the plasticity framework is often enhanced to incorporate the effects of degradation of mechanical properties (i.e., chemo-plasticity), creep (i.e., visco-plasticity), etc. In these cases, the time will explicitly enter the constitutive framework and will be inherently linked with the numerical integration scheme.

With this in mind, let us summarize the basic equations that govern the plasticity framework. The flow and evolution laws take the form

$$\begin{cases} \dot{\boldsymbol{\varepsilon}}^p = \dot{\lambda}\, \partial_{\sigma} \psi(\boldsymbol{\sigma}) \\ \dot{\boldsymbol{\alpha}} = \dot{\lambda}\, \boldsymbol{h}(\boldsymbol{\sigma}, \boldsymbol{\alpha}) \end{cases} \tag{5.15}$$

where $\psi(\boldsymbol{\sigma})$ is the plastic potential and ∂_σ is a differential operator (i.e. partial derivative with respect to $\boldsymbol{\sigma}$). The function $\boldsymbol{h}(\boldsymbol{\sigma}, \boldsymbol{\alpha})$ describes the evolution of internal variables and its form is consistent with the type of hardening law used. For the deviatoric hardening, for example, there is $\alpha = \varepsilon^p_q$, so that

$$\dot{\varepsilon}^p_q = \dot{\lambda}\frac{\partial \psi}{\partial q} \quad \rightarrow h(\boldsymbol{\sigma}, \varepsilon^p_q) = \frac{\partial \psi}{\partial q} \tag{5.16}$$

and $h(\boldsymbol{\sigma}, \varepsilon^p_q)$ is a simple scalar-valued function.

The general evolution problem (5.15) is subject to the initial conditions

$$\{\boldsymbol{\varepsilon}, \boldsymbol{\varepsilon}^p, \boldsymbol{\alpha}\}_{t=t_n} = \{\boldsymbol{\varepsilon}_n, \boldsymbol{\varepsilon}^p_n, \boldsymbol{\alpha}_n\} \tag{5.17}$$

and it is constrained by the following Kuhn-Tucker complementary conditions

$$f(\boldsymbol{\sigma},\boldsymbol{\alpha}) \leq 0, \ \ \dot{\lambda} \geq 0, \ \ \dot{\lambda} f(\boldsymbol{\sigma},\boldsymbol{\alpha}) = 0 \tag{5.18}$$

According to (5.18), the loading/unloading criterion is defined in the following general form

$$\begin{cases} \text{If} \quad f(\boldsymbol{\sigma},\boldsymbol{\alpha}) = 0 \text{ and } \dot{f}(\boldsymbol{\sigma},\boldsymbol{\alpha}) = 0 \text{ then} \\ \qquad \dot{\lambda} \geq 0 \\ \text{else} \\ \qquad \dot{\lambda} = 0 \end{cases} \tag{5.19}$$

Thus, in the plastic regime there is $\dot{f} = 0$ (stress remains on the current loading surface) and the plastic multiplier is positive, while in the elastic regime f remains negative and the plastic multiplier is zero.

The first step in the numerical integration is to transform the above defined evolution problem into a discrete constrained optimization problem for which an appropriate algorithmic procedure can be developed.

5.4.2 Notion of closest point projection

Applying the backward Euler scheme to (5.15), the problem can be formulated in a discrete sense as

$$\begin{cases} \boldsymbol{\varepsilon}_{n+1} = \boldsymbol{\varepsilon}_n + \Delta\boldsymbol{\varepsilon}_n \\ \boldsymbol{\varepsilon}^p_{n+1} = \boldsymbol{\varepsilon}^p_n + \Delta\lambda \partial_\sigma \psi(\boldsymbol{\sigma}_{n+1}) \\ \boldsymbol{\alpha}_{n+1} = \boldsymbol{\alpha}_n + \Delta\lambda\, \boldsymbol{h}(\boldsymbol{\sigma}_{n+1}, \boldsymbol{\alpha}_{n+1}) \\ \boldsymbol{\sigma}_{n+1} = [D^e](\boldsymbol{\varepsilon}_{n+1} - \boldsymbol{\varepsilon}^p_{n+1}) \end{cases} \tag{5.20}$$

with the Kuhn-Tucker conditions given by

$$\begin{cases} f(\boldsymbol{\sigma}_{n+1}, \boldsymbol{\alpha}_{n+1}) \leq 0 \\ \Delta\lambda \geq 0 \\ \Delta\lambda f(\boldsymbol{\sigma}_{n+1}, \boldsymbol{\alpha}_{n+1}) = 0 \end{cases} \tag{5.21}$$

As mentioned earlier in Section 5.2.2, an essential step in the solution of this discrete system is the selection of the trial stress state. In most cases, this state is identified with the purely elastic solution, i.e.

$$\begin{cases} \boldsymbol{\varepsilon}^{e\,(trial)}_{n+1} = \boldsymbol{\varepsilon}_{n+1} - \boldsymbol{\varepsilon}^p_n \\ \boldsymbol{\sigma}^{trial}_{n+1} = [D^e](\boldsymbol{\varepsilon}_{n+1} - \boldsymbol{\varepsilon}^p_n) \\ \boldsymbol{\alpha}^{trial}_{n+1} = \boldsymbol{\alpha}_n \\ f^{trial}_{n+1} = f(\boldsymbol{\sigma}^{trial}_{n+1}, \boldsymbol{\alpha}^{trial}_{n+1}) \end{cases} \tag{5.22}$$

The loading criterion (5.19) should also be expressed in an algorithmic form by invoking the trial stress state (5.22). Using the discrete constraints (5.21), Simo and Hughes [39] have shown that if the domain enclosed by the current loading surface is convex then $f_{n+1}^{trial} \geq f_{n+1}$. In this case, the loading criterion can be stated as

$$\begin{cases} f_{n+1}^{trial} \leq 0 \Rightarrow \text{ elastic step, } \Delta\lambda = 0 \\ f_{n+1}^{trial} > 0 \Rightarrow \text{ plastic step, } \Delta\lambda > 0 \end{cases} \tag{5.23}$$

Note that in case of an active loading process, the trial state is outside the elastic domain, i.e. that enclosed by the current loading surface. Therefore, from the computational point of view, a nonlinear optimization problem needs to be solved that consists of returning the trial state back to the boundary of the elastic domain. The optimum solution requires the evaluation of the closest distance, in the sense of an energy norm, between the point of trial state and a convex set (elastic domain).

Based on minimization in the energy norm, Simo and Hughes [39] have shown that in case of associated perfect-plasticity, the solution $\boldsymbol{\varepsilon}_{n+1}^{e} = \boldsymbol{\varepsilon}_{n+1} - \boldsymbol{\varepsilon}_{n+1}^{p}$ is the closest point projection of the trial state $\boldsymbol{\varepsilon}_{n+1}^{e\,(trial)}$ onto the failure surface in strain space. Similarly, the solution $\boldsymbol{\sigma}_{n+1} = \boldsymbol{\sigma}_{n+1}^{trial} - \Delta\lambda[D^e]\partial_\sigma \psi(\boldsymbol{\sigma}_{n+1})$ is the closest point projection of the trial state $\boldsymbol{\sigma}_{n+1}^{trial}$ onto the failure surface in stress space, Figure 5.11. For associated strain-hardening plasticity it can been proven [39], with some restrictions imposed on the type of hardening law, that the actual solution $\{\boldsymbol{\sigma}_{n+1}, \boldsymbol{\alpha}_{n+1}\}$ is the closest point projection of the trial state $\{\boldsymbol{\sigma}_{n+1}^{trial}, \boldsymbol{\alpha}_n\}$ onto the boundary of convex elastic domain. In the general case, however, the exact solution is not available, so that an approximate one is sought using an iterative procedure. It is noted that the notion of the closest point projection cannot be generalized to non-associated plasticity either.

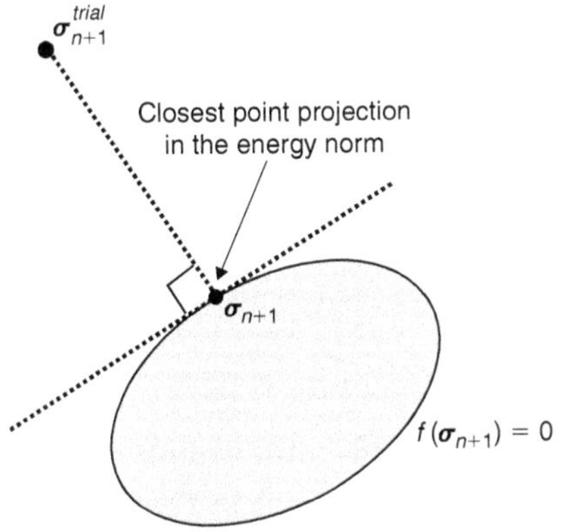

Figure 5.11 Geometrical representation of the closest point projection in the case of associated perfect-plasticity

5.4.3 Return-mapping algorithms

The numerical solution of the above nonlinear optimization problem is based on a two-step algorithm that includes

- an elastic trial predictor defined by (5.22);
- a plastic corrector which allows to return the trial state from a point outside the elastic domain back onto the loading surface.

This two-step algorithm emerges from an elastic-plastic operator split procedure which can be expressed in the following additive form (cf. Refs. [39,40])

$$\underset{\text{Total}}{\left\{\begin{array}{l}\dot{\boldsymbol{\varepsilon}}\\ \dot{\boldsymbol{\varepsilon}}^p=\dot{\lambda}\partial_\sigma\psi(\boldsymbol{\sigma})\\ \dot{\boldsymbol{\alpha}}=\dot{\lambda}h(\boldsymbol{\sigma},\boldsymbol{\alpha})\end{array}\right\}}=\underset{\text{Elastic predictor}}{\left\{\begin{array}{l}\dot{\boldsymbol{\varepsilon}}\\ \dot{\boldsymbol{\varepsilon}}^p=0\\ \dot{\boldsymbol{\alpha}}=0\end{array}\right\}}+\underset{\text{Plastic corrector}}{\left\{\begin{array}{l}\dot{\boldsymbol{\varepsilon}}=0\\ \dot{\boldsymbol{\varepsilon}}^p=\dot{\lambda}\partial_\sigma\psi(\boldsymbol{\sigma})\\ \dot{\boldsymbol{\alpha}}=\dot{\lambda}h(\boldsymbol{\sigma},\boldsymbol{\alpha})\end{array}\right\}} \tag{5.24}$$

The numerical process incorporates the following scheme. First, the elastic predictor problem is solved for the initial conditions (5.17). This generates an elastic trial state which, for an active loading process, is then taken as the initial condition for the solution of the plastic corrector problem. The objective of the second step is to restore the plastic consistency by "returning" the trial state to the current loading surface. The solution procedure for both these steps is described below.

(i) Evaluation of elastic predictor

The solution for the elastic predictor at time step t_n is easily obtained by taking

$$\left\{\begin{array}{l}\boldsymbol{\varepsilon}_{n+1}=\boldsymbol{\varepsilon}_n+\Delta\boldsymbol{\varepsilon}\\ \boldsymbol{\varepsilon}^p_{n+1}=\boldsymbol{\varepsilon}^p_n\\ \boldsymbol{\alpha}^{trial}_{n+1}=\boldsymbol{\alpha}_n\end{array}\right. \tag{5.25}$$

and specifying the trial state as

$$\boldsymbol{\sigma}^{trial}_{n+1}=[D^e](\boldsymbol{\varepsilon}_{n+1}-\boldsymbol{\varepsilon}^p_n) \tag{5.26}$$

The loading criterion is then checked according to eq.(5.23), i.e.

$$\left\{\begin{array}{l}f(\boldsymbol{\sigma}^{trial}_{n+1},\boldsymbol{\alpha}^{trial}_{n+1})\le 0\Rightarrow \text{elastic step (the trial state is the final solution)}\\ f(\boldsymbol{\sigma}^{trial}_{n+1},\boldsymbol{\alpha}^{trial}_{n+1})> 0\Rightarrow \text{plastic corrector problem to be solved}\end{array}\right. \tag{5.27}$$

(ii) Plastic corrector; cutting-plane algorithm

Clearly, the plastic corrector phase is the most complex one in the operator split algorithm. It represents a nonlinear optimization problem which cannot, in general, be solved explicitly. Only an iterative solution can be found using appropriate return-mapping algorithms. In the classical von Mises plasticity, the radial return-mapping algorithm is typically used to find the closest point projection of the trial state onto the yield surface. For more complex plasticity models though, return-mapping algorithms had not been developed until the mid 1980's. General algorithms, which employ implicit iterative procedure, are complex in numerical implementation as they require the gradients of the flow/hardening rules to be evaluated (cf. Ref. [39]). Such a task may become exceedingly laborious for most constitutive models. Therefore, in what follows, a simplified return-mapping algorithm based on an explicit iterative procedure is outlined. This algorithm, referred to as the cutting-plane algorithm, was proposed by Simo and Ortiz [41,42] and it is commonly used in the numerical analysis of engineering problems.

In order to develop an iterative procedure, the plastic corrector problem in (5.24) is reformulated using backward Euler scheme, as

$$\begin{cases} \boldsymbol{\varepsilon}^p_{n+1} = \boldsymbol{\varepsilon}^p_n + \Delta\lambda \partial_\sigma \psi\{[D^e](\boldsymbol{\varepsilon}_{n+1} - \boldsymbol{\varepsilon}^p_n)\} \\ \boldsymbol{\alpha}_{n+1} = \boldsymbol{\alpha}_n + \Delta\lambda h(\boldsymbol{\sigma}_{n+1}, \boldsymbol{\alpha}_{n+1}) \end{cases} \tag{5.28}$$

Note that for any $t \in [t_n, t_n + \Delta t]$ there is $\boldsymbol{\varepsilon}^p = \boldsymbol{\varepsilon}^p(\Delta\lambda), \boldsymbol{\alpha} = \boldsymbol{\alpha}(\Delta\lambda)$, i.e. both variables are functions of the plastic multiplier $\Delta\lambda$. Taking the derivatives of both sides of the expressions (5.28) with respect to $\Delta\lambda$, one obtains

$$\begin{cases} \dfrac{d\boldsymbol{\varepsilon}^p(\Delta\lambda)}{d\Delta\lambda} = \partial_\sigma \psi\{[D^e](\boldsymbol{\varepsilon}_{n+1} - \boldsymbol{\varepsilon}^p(\Delta\lambda))\} \\ \dfrac{d\boldsymbol{\alpha}(\Delta\lambda)}{d\Delta\lambda} = h\{[D^e](\boldsymbol{\varepsilon}_{n+1} - \boldsymbol{\varepsilon}^p(\Delta\lambda)), \boldsymbol{\alpha}(\Delta\lambda)\} \end{cases} \tag{5.29}$$

subject to the initial conditions

$$\{\boldsymbol{\varepsilon}^p(\Delta\lambda), \boldsymbol{\alpha}(\Delta\lambda)\}|_{\Delta\lambda=0} = \{\boldsymbol{\varepsilon}^p_n, \boldsymbol{\alpha}_n\} \tag{5.30}$$

Using now the elastic stress-strain relation, the stress state is defined as

$$\boldsymbol{\sigma}(\Delta\lambda) = [D^e](\boldsymbol{\varepsilon}_{n+1} - \boldsymbol{\varepsilon}^p(\Delta\lambda)) \tag{5.31}$$

Substituting (5.31) in (5.29) yields

$$\begin{cases} \dfrac{d\boldsymbol{\sigma}(\Delta\lambda)}{d\Delta\lambda} = -[D^e]\dfrac{d\boldsymbol{\varepsilon}^p(\Delta\lambda)}{d\Delta\lambda} = -[D^e]\,\partial_\sigma \psi\{\boldsymbol{\sigma}(\Delta\lambda)\} \\ \dfrac{d\boldsymbol{\alpha}(\Delta\lambda)}{d\Delta\lambda} = h\{\boldsymbol{\sigma}(\Delta\lambda), \boldsymbol{\alpha}(\Delta\lambda)\} \end{cases} \tag{5.32}$$

with the initial conditions

$$\{\boldsymbol{\sigma}(\Delta\lambda), \boldsymbol{\alpha}(\Delta\lambda)\}|_{\Delta\lambda=0} = \{\boldsymbol{\sigma}^{trial}_{n+1}, \boldsymbol{\alpha}_n\} \tag{5.33}$$

The solution of the plastic corrector problem defined by (5.32) is a continuous curve $\Delta\lambda \in R_+ \to [\boldsymbol{\sigma}(\Delta\lambda), \boldsymbol{\alpha}(\Delta\lambda)]$ that starts from the initial conditions given by (5.33). The consistency condition is recovered by determining the intersection of this curve with

Table 5.4 Iterative procedure of the cutting-plane return-mapping algorithm for plastic corrector problem.

1. Initialize $k=0$, $\boldsymbol{\varepsilon}^{p(0)}_{n+1}=\boldsymbol{\varepsilon}^{p}_{n}$, $\boldsymbol{\alpha}^{(0)}_{n+1}=\boldsymbol{\alpha}_{n}$, $\Delta\lambda^{(0)}_{n+1}=0$
2. Compute:

$$\begin{cases} \boldsymbol{\sigma}^{(k)}_{n+1}=[D^e](\boldsymbol{\varepsilon}_{n+1}-\boldsymbol{\varepsilon}^{p(k)}_{n+1}) \\ \boldsymbol{h}^{(k)}_{n+1}=\boldsymbol{h}[\boldsymbol{\sigma}^{(k)}_{n+1},\boldsymbol{\alpha}^{(k)}_{n+1}] \\ f^{(k)}_{n+1}=f[\boldsymbol{\sigma}^{(k)}_{n+1},\boldsymbol{\alpha}^{(k)}_{n+1}] \end{cases}$$

If $f^{(k)}_{n+1}\leq Tol.$ then EXIT
else

3. Compute the variation of plastic multiplier:

$$\Delta^2\lambda^{(k+1)}_{n+1}=\frac{f^{(k)}_{n+1}}{\partial_\sigma f^{(k)}_{n+1}[D^e]\partial_\sigma\psi^{(k)}_{n+1}-\partial_\alpha f^{(k)}_{n+1}\boldsymbol{h}^{(k)}_{n+1}}$$

4. Update state variables and the plastic multiplier:

$$\boldsymbol{\varepsilon}^{p,(k+1)}_{n+1}=\boldsymbol{\varepsilon}^{p,(k)}_{n+1}+\Delta^2\lambda^{(k+1)}_{n+1}\partial_\sigma\psi^{(k)}_{n+1}$$
$$\boldsymbol{\alpha}^{(k+1)}_{n+1}=\boldsymbol{\alpha}^{(k)}_{n+1}+\Delta^2\lambda^{(k+1)}_{n+1}\boldsymbol{h}^{(k)}_{n+1}$$
$$\Delta\lambda^{(k+1)}_{n+1}=\Delta\lambda^{(k)}_{n+1}+\Delta^2\lambda^{(k+1)}_{n+1}$$

Set $k=k+1$; go to (2)

the boundary of the elastic domain. Thus, the solution of (5.33) is equivalent to that of the following problem

$$\text{find } \Delta\lambda\in R_+ \text{ such that: } \bar{f}(\Delta\lambda)=f\{\boldsymbol{\sigma}(\Delta\lambda),\boldsymbol{\alpha}(\Delta\lambda)\}=0 \tag{5.34}$$

In order to solve (5.34) an iterative procedure is employed starting with $\Delta\lambda^{(0)}=0$. Let $\Delta\lambda^{(k)}$ be the value of $\Delta\lambda$ at the kth iteration. The constraint (5.34) can be linearized as

$$\begin{aligned} &f\{\boldsymbol{\sigma}(\Delta\lambda),\boldsymbol{\alpha}(\Delta\lambda)\}=f\{\boldsymbol{\sigma}(\Delta\lambda^{(k)}),\boldsymbol{\alpha}(\Delta\lambda^{(k)})\}\\ &-\partial_\sigma f\{\boldsymbol{\sigma}(\Delta\lambda^{(k)}),\boldsymbol{\alpha}(\Delta\lambda^{(k)})\}[D^e]\partial_\sigma\psi\{\boldsymbol{\sigma}(\Delta\lambda^{(k)})\}\Delta^2\lambda^{(k+1)}\\ &+\partial_\alpha f\{\boldsymbol{\sigma}(\Delta\lambda^{(k)}),\boldsymbol{\alpha}(\Delta\lambda^{(k)})\}\boldsymbol{h}\{\boldsymbol{\sigma}(\Delta\lambda^{(k)}),\boldsymbol{\alpha}(\Delta\lambda^{(k)})\}\Delta^2\lambda^{(k+1)}=0 \end{aligned} \tag{5.35}$$

This, in turn, leads to an expression for the variation of the plastic multiplier $\Delta^2\lambda^{(k+1)}$, which takes the form

$$\Delta^2\lambda^{(k+1)}=\frac{f^{(k)}}{\partial_\sigma f^{(k)}[D^e]\partial_\sigma\psi^{(k)}-\partial_\alpha f^{(k)}\boldsymbol{h}^{(k)}}=\frac{f^{(k)}}{H_e+H_p} \tag{5.36}$$

where the specific expressions for the hardening moduli H_e, H_p are provided in Chapter 3 for both the volumetric and deviatoric hardening rules. The cumulative value of the plastic multiplier becomes

$$\Delta\lambda^{(k+1)}=\Delta\lambda^{(k)}+\Delta^2\lambda^{(k+1)} \tag{5.37}$$

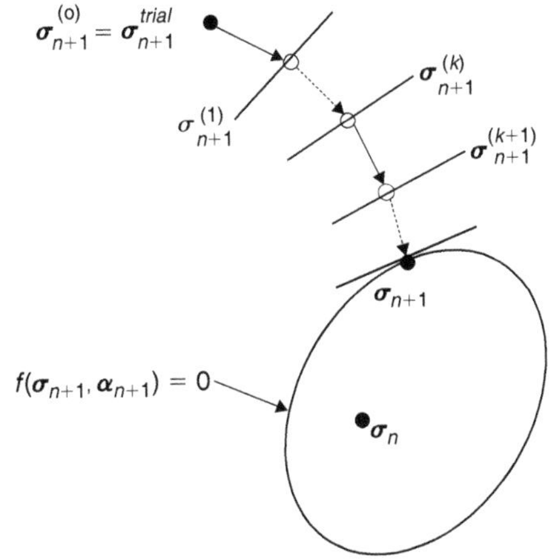

Figure 5.12 Geometrical illustration of the iterative procedure in the cutting-plane algorithm for plastic corrector problem

Given the above iterative solution of the linearized problem, the cutting-plane algorithm for plastic corrector may be formulated in three main steps:

1. Assume $f_{n+1}^{trial} > 0$ and then $\Delta\lambda_{n+1} > 0$; integrate explicitly the return-mapping equations (5.32) over an interval of length $\Delta\lambda$ yet unknown;
2. using the linearized form (5.35), solve for the variation $\Delta^2\lambda$ from the relation (5.36);
3. update $\Delta\lambda$ with (5.37), together with all state variables, and check the condition (5.34); return to step (1) if the latter is not satisfied.

Note that this procedure requires only functional evaluations without computing the gradients of flow rule and hardening laws.

The cutting-plane return-mapping algorithm is summarized in Table 5.4. A geometrical illustration of the iterative procedure is provided in Figure 5.12.

Chapter 6

Introduction to limit analysis

The general topic of plasticity in geomechanics cannot be completed without making at least a brief reference to the approach based on limit analysis. The present chapter provides a concise introduction to this subject.

Irreversibility and path-dependence require, in general, an incremental step-by-step analysis that follows the history of loading. For many practical problems, however, it is often satisfactory to define the critical load associated with impending collapse of the structure. This can be accomplished by invoking a simplified approach in which the material is idealized as rigid-perfectly plastic. In this case, closed-form solutions can be found for a limited number of problems. Alternatively, lower and upper bounds can be established on the collapse load by employing the limit theorems. In the latter case, close approximations to the limit load and sometimes even exact solutions can be obtained for a variety of practical geotechnical problems.

The upper and lower bound theorems were proven as early as 1936 [43], however their full potential was recognized much later, i.e. in 1950's [44, 45]. Since that time, there have been an enormous number of applications in the broad area of solid mechanics, including both structural as well as geomechanics (see, e.g. [46–48]). The methodology is particularly attractive since the theorems of limit analysis can be rigorously applied to *elastic* – perfectly plastic bodies as well. This stems from positive definiteness of the elastic complimentary energy combined with Drucker's inequality (1.38). The formal proof is given, for example, in Refs.[44] and [49]. This chapter provides a brief review of the fundamentals of the limit analysis. The emphasis here is on the formulation of the problem and its subsequent illustration, the latter involving simple applications in the area of geotechnical engineering.

6.1 FORMULATION OF LOWER AND UPPER BOUND THEOREMS

Consider a body of volume V and the surface area S. Assume that the state of stress inside the body, σ_{ij}, satisfies equilibrium equations and corresponds to traction $T_i = \sigma_{ij} n_j$ acting along the surface S_T, Figure 6.1a. Imagine now a different stress distribution σ^*_{ij} satisfying static boundary conditions on S_T (i.e. $\sigma^*_{ij} n_j = T^*_i = T_i$) but

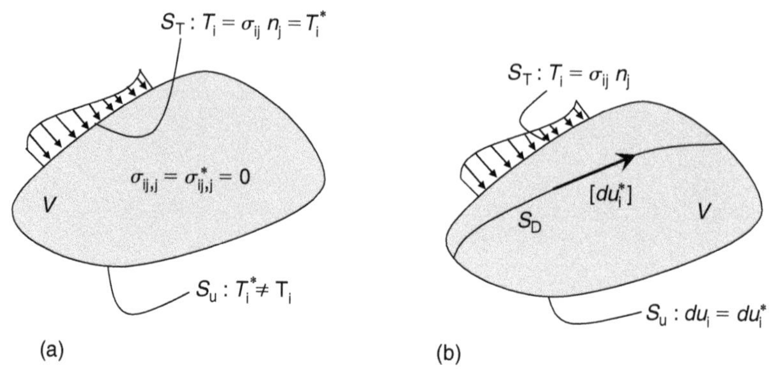

Figure 6.1 Definition of the problem; (a) static and (b) kinematic approach

giving rise to different surface traction on S_u ($T_i^* \neq T_i$ on S_u). Thus,

$$\sigma_{ij}\!: \quad T_i = \sigma_{ij} n_j \text{ (on } S_T\text{)}; \quad \sigma_{ij,j} = 0 \text{ (in } V\text{)}$$

$$\sigma_{ij}^*\!: \quad T_i^* = \sigma_{ij}^* n_j = T_i \text{ (on } S_T\text{)}; \quad \sigma_{ij,j}^* = 0 \text{ (in } V\text{)}$$

$$T_i^* \neq T_i \text{ (on } S_u\text{)}$$

Assume now that we impose a velocity $\dot{u}_i$ on S_u. Since both σ_{ij} and σ_{ij}^* satisfy equilibrium, we can write the principle of virtual work for both these fields, i.e.

$$\int_S T_i \, du_i \, dS = \int_V \sigma_{ij} \, d\varepsilon_{ij} \, dV; \quad \int_S T_i^* \, du_i \, dS = \int_V \sigma_{ij}^* \, d\varepsilon_{ij} \, dV \tag{6.1}$$

Subtracting the above equations, one obtains

$$\int_S (T_i - T_i^*) \, du_i \, dS = \int_V (\sigma_{ij} - \sigma_{ij}^*) \, d\varepsilon_{ij} \, dV \tag{6.2}$$

or, since $T_i = T_i^*$ on S_T

$$\int_{S_u} (T_i - T_i^*) \, du_i \, dS_u = \int_V (\sigma_{ij} - \sigma_{ij}^*) \, d\varepsilon_{ij} \, dV \tag{6.3}$$

Assume, at this point, that the material is *rigid-perfectly plastic* and that both σ_{ij} and σ_{ij}^* satisfy the failure criterion. Assume further that the material is stable in Drucker's sense, i.e. the convexity and the normality conditions are satisfied. Since σ_{ij} represents the actual stress field, the postulate of maximum plastic work ensures that

$$\int_V (\sigma_{ij} - \sigma_{ij}^*) \, d\varepsilon_{ij} \, dV \geq 0 \tag{6.4}$$

Thus,

$$\int_{S_u} T_i \, du_i \, dS_u \geq \int_{S_u} T_i^* \, du_i \, dS_u \tag{6.5}$$

Inequality (6.5) states that the work performed by actual traction (T_i) is larger or equal to that done by a fictitious traction (T_i^*) on the same virtual displacement rate. Thus, this assertion provides a *lower bound* assessment on the actual rate of work done. The stress field σ_{ij}^* is referred to as *statically and plastically admissible*, which implies that it satisfies equilibrium, boundary conditions and does not violate the failure criterion.

In order to provide the *upper bound* estimate, introduce a fictitious velocity field $\dot{u}_i^*$ that satisfies the kinematic boundary conditions, i.e. $\dot{u}_i = \dot{u}_i^*$ *on* S_u. Such a velocity field is referred to as *kinematically admissible* and may, in general, have a finite set of discontinuities $[\dot{u}_i^*] = [(\dot{u}_i^*)^+ - (\dot{u}_i^*)^-]$ which occur along pre-defined surfaces S_D (Figure 6.1b). The formulation of the problem involves writing the principle of virtual work for the velocity field $\dot{u}_i^*$ and the actual stress field σ_{ij}, i.e.

$$\int_S T_i \, du_i^* \, dS = \int_V \sigma_{ij} \, d\varepsilon_{ij}^* \, dV + \sum \int_{S_D} t_i [du_i^*] \, dS_D \tag{6.6}$$

where t_i is the traction vector along the discontinuity surface and $d\varepsilon_{ij}^*$ is the plastic strain increment derived from du_i^*. Let σ_{ij}^* be the stress field giving rise to $d\varepsilon_{ij}^*$. In this case, the principle of maximum plastic work takes the form

$$\int_V (\sigma_{ij}^* - \sigma_{ij}) \, d\varepsilon_{ij}^* \, dV \geq 0 \tag{6.7}$$

Substituting inequality (6.7) into the statement of the virtual work principle, yields

$$\int_S T_i \, du_i^* \, dS \leq \int_V \sigma_{ij}^* \, d\varepsilon_{ij}^* \, dV + \sum \int_{S_D} t_i [du_i^*] \, dS_D \tag{6.8}$$

Now, since the term on the left-hand side satisfies

$$\int_S T_i \, du_i^* \, dS = \int_{S_u} T_i \, du_i \, dS_u + \int_{S_T} T_i \, du_i^* \, dS_T \tag{6.9}$$

the following inequality is obtained

$$\int_{S_u} T_i \, du_i \, dS_u + \int_{S_T} T_i \, du_i^* \, dS_T \leq \int_V \sigma_{ij}^* \, d\varepsilon_{ij}^* \, dV + \sum \int_{S_D} t_i [du_i^*] \, dS_D \tag{6.10}$$

Here, the left-hand side represents the rate of work done by external agencies, while the term on the right gives the rate of internal energy dissipation. Clearly, the equality sign holds only in the case when the kinematically admissible velocity field $\dot{u}_i^*$ coincides with the actual one, i.e. $\dot{u}_i$. Note that by combining inequalities (6.5) and (6.10) the upper and lower bound estimates for the actual rate of work can be obtained, i.e.

$$\int_V \sigma_{ij}^* \, d\varepsilon_{ij}^* \, dV + \sum \int_{S_D} t_i[du_i^*] \, dS_D - \int_{S_T} T_i \, du_i^* \, dS_T \geq \int_{S_u} T_i \, du_i \, dS_u$$
$$\geq \int_{S_u} T_i^* \, du_i \, dS_u \tag{6.11}$$

In this case, the left-hand side employs a kinematically admissible velocity field, while the right-hand side incorporates a statically and plastically admissible stress field. The inequalities (6.11) indicate a direct method of approximating the limit load in case when the velocities are prescribed over S_u. Simple and important results can also be obtained for problems that involve a proportional loading acting over S_T.

Consider the case when the surface load increases in proportion to a parameter $\beta > 0$. To find the lower bound of this load factor, consider an admissible stress field σ_{ij}^* which satisfies somewhat modified boundary conditions along S_T, i.e.

$$T_i = \sigma_{ij} n_j = \beta T_i^0; \quad T_i^* = \sigma_{ij}^* n_j = \beta_s T_i^0 \text{ (on } S_T) \tag{6.12}$$

where β_s is the static loading factor associated with the stress field σ_{ij}^*. Employing these conditions, the inequality (6.5) can be expressed as

$$\beta \int_S T_i^0 \, du_i \, dS \geq \beta_s \int_S T_i^0 \, du_i \, dS \quad \Rightarrow \quad \beta \geq \beta_s \tag{6.13}$$

Thus, the statement of the *lower bound theorem* can be re-phrased as follows:

If a statically and plastically admissible stress field σ_{ij}^* can be found for a body under a given set of loads, the body will either not collapse or may be just on the verge of collapse.

The upper bound on β can be obtained from consideration of inequality (6.10). Assume that T_i^0 represents again a certain fixed distribution of traction over S_T, while the boundary S_u is kinematically constrained, so that the velocities are zero. Thus,

$$T_i = \beta T_i^0 \text{ (on } S_T); \quad du_i = du_i^* = 0 \text{ (on } S_u) \tag{6.14}$$

In this case,

$$\beta \int_{S_T} T_i^0 \, du_i^* \, dS_T \leq \int_V \sigma_{ij}^* \, d\varepsilon_{ij}^* \, dV + \sum \int_{S_D} t_i[du_i^*] \, dS_D \tag{6.15}$$

so that

$$\beta \leq \beta_k = \frac{\int_V \sigma_{ij}^* \, d\varepsilon_{ij}^* \, dV + \sum \int_{S_D} t_i[du_i^*] \, dS_D}{\int_{S_T} T_i^0 \, du_i^* \, dS_T} \tag{6.16}$$

where β_k is the kinematic loading factor associated with the virtual velocity field $\dot{u}_i^*$. Thus, the *upper bound theorem* can be re-stated as follows:

If a kinematically admissible velocity field $\dot{u}_i^*$ can be found for a body under a given set of loads, then the body must be on the verge of collapse or must have already collapsed.

Note that the work done by surface traction on the kinematically admissible velocities, which appears in the denominator of (6.16), must be positive.

Let us focus our attention on specification of the energy dissipated along the discontinuity surfaces S_D, which is represented by the last term in the virtual work expression (6.6). Consider first *Tresca* material, for which the failure function F assumes the form

$$F = t_i s_i - c = 0; \qquad t_i = \sigma_{ij} n_j \tag{6.17}$$

Here, n_i is the unit normal to the discontinuity surface S_D, c is the shear strength and s_i specifies the direction of the resultant shear traction. The latter is defined as

$$s_i = \frac{t_i^s}{|t_i^s|}; \quad t_i^s = t_i - t_i^n = t_i - n_i n_k t_k = (\delta_{ik} - n_i n_k)\, t_k \tag{6.18}$$

Note that $t_i s_i = |t_i^s| \geq 0$. Also, $t_i^s n_i = 0$ which, in turn, implies that s_i is orthogonal to n_i, i.e. $s_i n_i = 0$.

The rate of energy dissipation can be evaluated as

$$D = t_i\,[du_i^*]; \quad [du_i^*] = [(du_i^*)^+ - (du_i^*)^-] \tag{6.19}$$

For a stable material, in Drucker's sense, an associated flow rule is required, so that

$$[du_i^*] \propto \frac{\partial F}{\partial t_i} \Rightarrow [du_i^*] = |du^*|\, s_i \tag{6.20}$$

where $|du^*| = |\,[du_i^*]\,|$ is the magnitude of $[du_i^*]$. Thus, in this case, the velocity discontinuity $[\dot{u}_i^*]$ is tangential to S_D, i.e. satisfies $[(\dot{u}_i^*)^+ - (\dot{u}_i^*)^-]\, n_i = 0$, and the rate of energy dissipation becomes

$$D = t_i\,|du^*|\, s_i = c\,|du^*| \tag{6.21}$$

For geomaterials, the interface behaviour is typically formulated by employing *Coulomb* criterion

$$F = t_i s_i + \mu\, t_i n_i - c = 0; \qquad \mu = \tan\phi \tag{6.22}$$

where ϕ is the angle of friction and c is now referred to as the cohesion. Note again that setting $\phi = 0$ in (6.22) leads to the former representation (6.17). For Coulomb material, the associated flow rule requires

$$[du_i^*] \propto \frac{\partial F}{\partial t_i} = s_i + \mu\, n_i \tag{6.23}$$

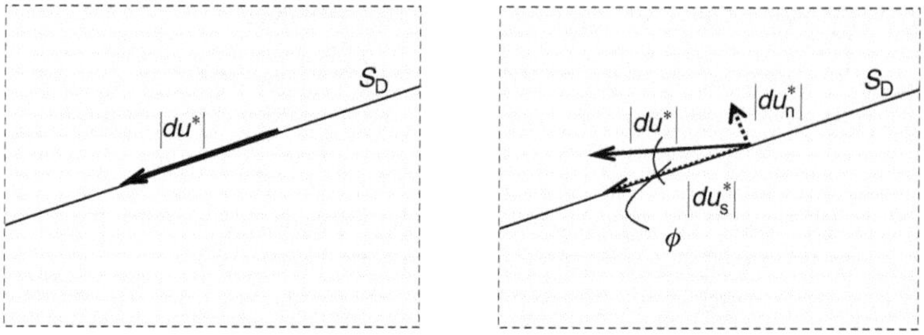

Figure 6.2 Direction of the resultant velocity discontinuity in Tresca and Coulomb material

so that

$$[du_i^*] = |du_s^*|\, s_i + |du_n^*|\, n_i; \qquad |du_n^*| = \mu\, |du_s^*| \tag{6.24}$$

Thus, the discontinuity in the tangential component of velocity $|du_s^*|$ requires a separation velocity $|du_n^*|$. The direction of the resultant velocity discontinuity is confined to the plane formed by the base vectors n_i, s_i and it deviates from s_i by the angle ϕ, Figure 6.2. The rate of energy dissipation becomes

$$\begin{aligned} D &= t_i\,[du_i^*] = t_i\, |du_s^*|\, s_i + t_i\, |du_n^*|\, n_i = (t_i s_i + \mu\, t_i n_i)\, |du_s^*| \\ &= c\, |du_s^*| = c\, |du^*| \cos\phi \end{aligned} \tag{6.25}$$

where $|du^*|$ is the magnitude of the resultant velocity discontinuity. It is noted that, according to (6.25), in a *cohesionless* material, satisfying an associated flow rule, there is $D = 0$, i.e. no energy is being dissipated.

6.2 EXAMPLES OF APPLICATIONS OF LIMIT THEOREMS IN GEOTECHNICAL ENGINEERING

In this section, some elementary examples of application of limit analysis, incorporating lower/upper bound theorems, are provided. It needs to be emphasized that the scope of presentation here is limited and focuses on a class of standard problems that illustrate the methodology involved. A far more comprehensive coverage is provided in a number of monographs that deal solely with the topic of limit analysis, such as those by Chen and co-workers [47, 48].

(i) Tresca material

Let us start with an indentation problem that is classical in the theory of plasticity. The material considered here is assumed to satisfy the Tresca criterion that, in terms of principal stress, takes the form

$$F = |\sigma_1 - \sigma_3| - 2c = 0 \tag{6.26}$$

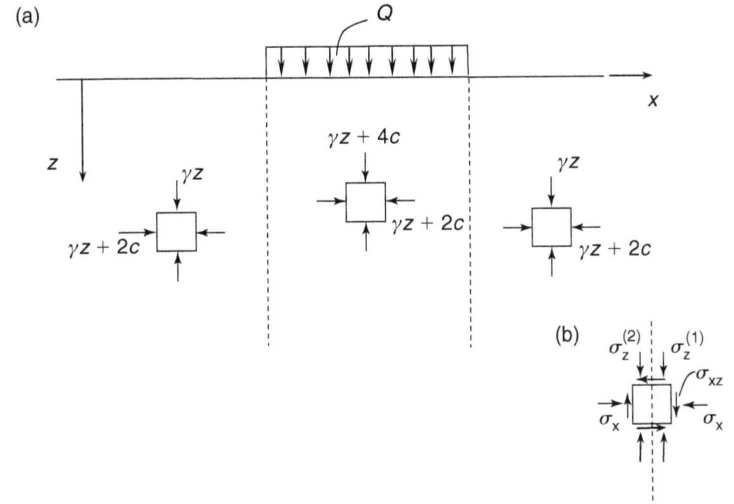

Figure 6.3 Lower bound solution; an elementary stress field with two planes of discontinuity ($Q_f \geq 4c$)

The latter can be obtained directly from Mohr-Coulomb representation (2.20), by setting $\phi = 0$.

It is noted that, in geotechnical practice, the Tresca criterion is often used to assess the stability of structures immediately after construction (undrained conditions). The failure condition (6.26) is then interpreted in the context of total, rather than effective, stress and the constant c is identified with the undrained shear strength. One can certainly argue that such approach is, in general, questionable; it is, however, quite extensively used in geotechnical design. Thus, the classical indentation problem, as stated earlier, may be perceived here as an assessment of bearing capacity of a shallow foundation under undrained constraint.

Consider first the lower bound approach. Figure 6.3a presents an elementary stress field that can be assumed in the first approximation. It incorporates two planes of stress discontinuity, as shown. Note that the equilibrium requires that along the discontinuity plane all stress components, except for normal stress parallel to the plane, remain continuous, Figure 6.3b. The solution is now constructed as follows. In the region away from the foundation, a uniform stress field is stipulated in which the vertical component is that due to gravity, i.e. $\sigma_z = \sigma_1 = -\gamma z$, where γ is the unit weight of the material. The horizontal component is chosen now such that the Tresca criterion (6.26) is satisfied; thus, $\sigma_x = \sigma_3 = -(\gamma z + 2c)$. Note that this stress field satisfies equilibrium, the boundary condition ($z = 0 \Rightarrow \sigma_z = 0$) and does not violate the failure criterion (i.e. $F = 0$), so that it is both statically and plastically admissible. In the region beneath the foundation the horizontal stress must remain continuous, i.e. $\sigma_x = -(\gamma z + 2c)$, while the vertical stress may again be chosen in such a way as to satisfy $F = 0$, eq.(6.26). To obtain the *maximum* value of the critical load, for the given stress field, we take $\sigma_z = \sigma_3 = -(\gamma z + 4c)$. Since $z = 0$ requires $\sigma_z = -Q$, there must be $Q = 4c$. Invoking now the lower bound theorem, eq.(6.13), the estimate for the failure load becomes $Q_f \geq 4c$.

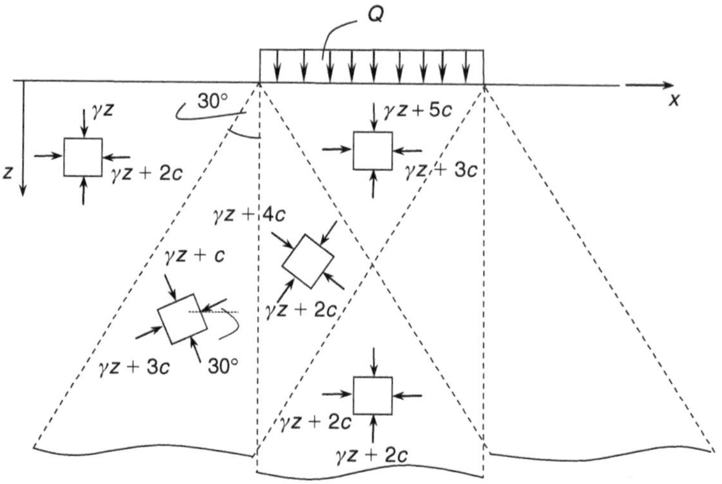

Figure 6.4 Lower bound solution; a more complex stress field yielding $Q_f \geq 5c$

By the virtue of the lower bound theorem, the higher the estimate for the failure load, the closer it is to the exact solution. With this in mind, consider a more complex stress field which now incorporates a set of inclined discontinuity planes, as shown in Figure 6.4. In this case, the same initial stress state $\sigma_z = \sigma_1 = -\gamma z$ and $\sigma_x = \sigma_3 = -(\gamma z + 2c)$ is assumed and it is subsequently transformed to a local coordinate system attached to the discontinuity plane. In this configuration, the normal and shear stresses must now remain continuous while the discontinuity in the component parallel to the plane may again be chosen such that $F = 0$. Note that in order to evaluate the admissible stress discontituity, the criterion (6.26) must now be expressed in terms of Cartesian, rather than principal stress components. The above procedure results in the stress field for which the principal stress magnitudes are $\sigma_1 = -(\gamma z + c)$; $\sigma_3 = -(\gamma z + 3c)$, as shown in Figure 6.4. Following the same methodology across subsequent discontinuity planes, the stress state beneath the foundation is established for which $\sigma_z = \sigma_3 = -(\gamma z + 5c)$. Again, since $z = 0$ requires $\sigma_z = -Q$, there must be $Q = 5c$, so that by the statement of the lower bound theorem, the estimate for the failure load now becomes $Q_f \geq 5c$. The latter is apparently more accurate than the one corresponding to the stress field in Figure 6.3.

Given the lower bound estimate, let us turn our attention now to the upper bound approach. In this case a kinematically admissible velocity field needs to be postulated, while the resulting estimate of the ultimate load would be considered as unsafe. Thus, the more accurate the kinematic mechanism, the lower the assessment of the failure load. In what follows, a sequence of kinematically admissible velocity fields is examined that result in a progressively lower estimate of the ultimate bearing capacity. The primary mechanisms considered here are shown in Figure 6.5. They all employ a set of rigid wedges that slide relative to each other and form a global kinematic mechanism on the macroscale. The details on the specification of the failure load, for each case considered, are provided in tables that are appended to Figures 6.6–6.9.

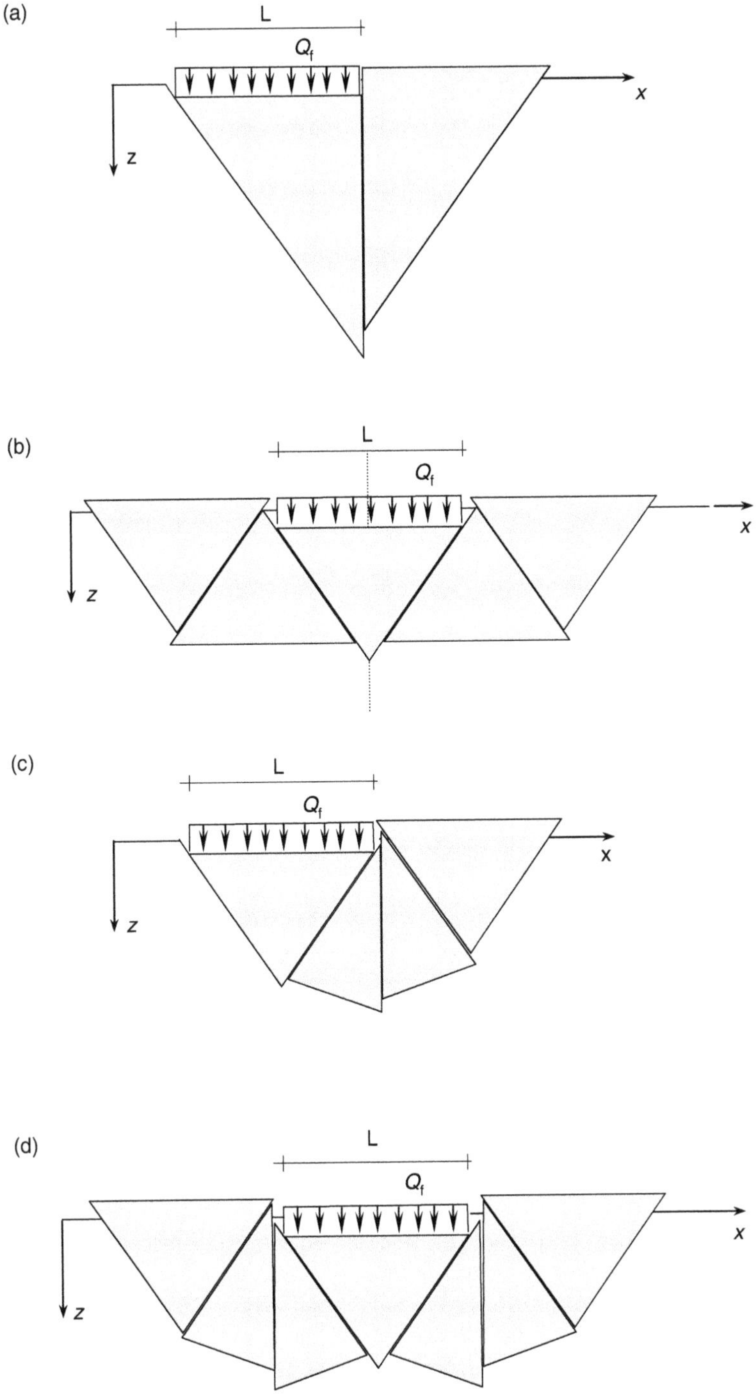

Figure 6.5 A sequence of the considered kinematically admissible velocity fields

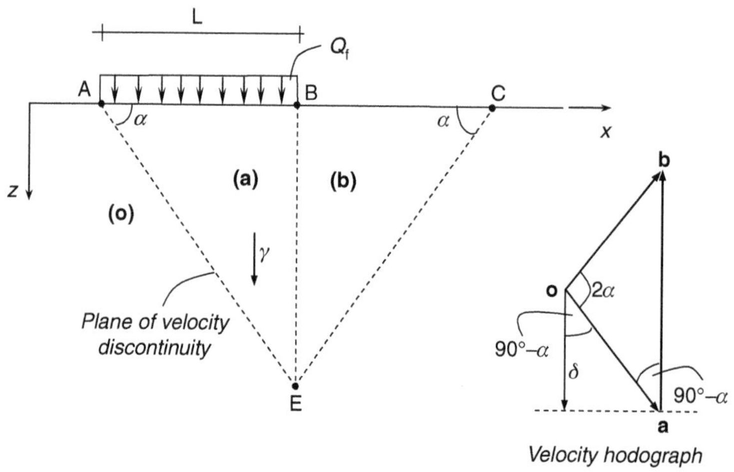

Discontinuity plane	Length	Velocity discontinuity	Dissipation rate
AE	$L/\cos\alpha$	$\lvert\mathbf{oa}\rvert=\delta/\sin\alpha$	$cL\delta/(\sin\alpha\cos\alpha)$
BE	$L\tan\alpha$	$\lvert\mathbf{ab}\rvert=2\delta$	$2cL\delta\tan\alpha$
CE	$L/\cos\alpha$	$\lvert\mathbf{ob}\rvert=\delta/\sin\alpha$	$cL\delta/(\sin\alpha\cos\alpha)$

Total dissipation rate:

$$D=2c\delta L\left(\frac{1+\sin^2\alpha}{\sin\alpha\cos\alpha}\right)$$

Rate of external work:

$$W=Q_f\,L\delta+\frac{1}{2}\gamma L|BE|\delta-\frac{1}{2}\gamma|CB||BE|\delta=Q_f\,L\delta$$

Upper bound theorem:

$$W\le D\Rightarrow Q_f\le\min_{\alpha}\left\{2c\left(\frac{1+\sin^2\alpha}{\sin\alpha\cos\alpha}\right)\right\}=5.66c\ (\text{for }\alpha=35.3°)$$

Figure 6.6 Upper bound solution for the mechanism in Figure 6.5a

Examine the first velocity field as identified in Figure 6.5. The mechanism depicted here incorporates two rigid wedges that slide along the discontinuity planes (Figure 6.6). Velocity in the region (o), away from the foundation, is assumed to be zero. The wedge (a) slides along AE thus forcing (b) to move up (along BE) and to the right (along EC). The corresponding velocity hodograph is shown in the adjacent figure. The foundation is given a constant velocity δ and, based on the geometry of the problem, the velocity discontinuities along all the planes are identified (see the table in Figure 6.6). By evaluating now the rate of external work and invoking the upper bound theorem, an estimate for the bearing capacity is found. For the mechanism shown the minimum value of $Q_f\le 5.66c$ is obtained for $\alpha=35.3°$. Thus, recalling the lower bound approximation of Figure 6.3, there is $5c\le Q_f\le 5.66c$.

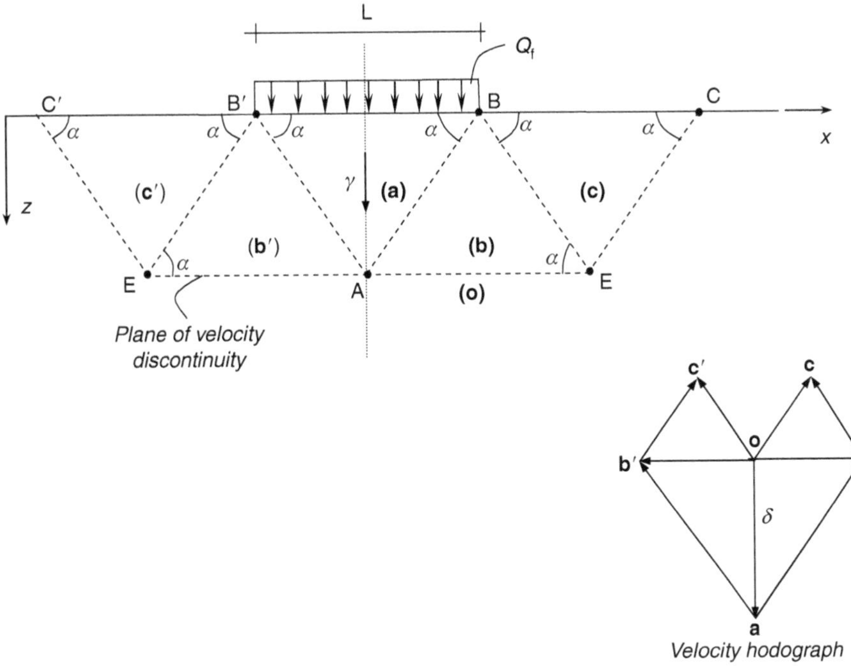

Discontinuity plane	*Length*	*Velocity discontinuity*
AB, AB′	$L/(2\cos\alpha)$	$\lvert\mathbf{ab}\rvert = \lvert\mathbf{ab}'\rvert = \delta/\sin\alpha$
BE, B′E′	$L/(2\cos\alpha)$	$\lvert\mathbf{bc}\rvert = \lvert\mathbf{b}'\mathbf{c}'\rvert = \delta/(2\sin\alpha)$
CE, C′E′	$L/(2\cos\alpha)$	$\lvert\mathbf{oc}\rvert = \lvert\mathbf{oc}'\rvert = \delta/(2\sin\alpha)$
AE, AE′	L	$\lvert\mathbf{ob}\rvert = \lvert\mathbf{ob}'\rvert = \delta/\tan\alpha$

Total dissipation rate:

$$D = 2cL\delta\left[\frac{1+\cos^2\alpha}{\sin\alpha\cos\alpha}\right]$$

*Rate of external work**:

$$W = Q_f L\delta$$

Upper bound theorem:

$$Q_f \le \min_{\alpha}\left\{2c\frac{1+\cos^2\alpha}{\sin\alpha\cos\alpha}\right\} \le 5.66\,c \text{ (for } \alpha = 54.7^\circ)$$

(*)Note: rate of work of body forces equals zero

Figure 6.7 Upper bound solution for the mechanism in Figure 6.5b

A similar methodology can be employed to estimate the bearing capacity based on the second mechanism of Figure 6.5. In this case, the wedge (**a**) moves downward with the velocity δ, while the wedges (**b**), (**c**) and (**b′**), (**c′**) move along the respective velocity discontinuity planes AE, EC and AE′, E′C′ (Figure 6.7). The assumed mechanism is now symmetric with respect to the centerline of the foundation and the estimated value

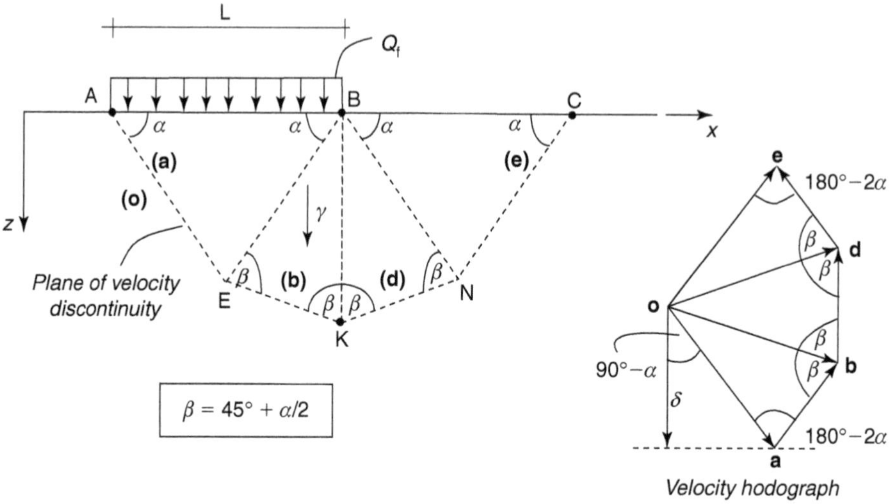

Discontinuity plane	*Length*	*Velocity discontinuity*
AE	$L/(2\cos\alpha)$	$\lvert\mathbf{oa}\rvert=\delta\sin\alpha$
BE	$L/(2\cos\alpha)$	$\lvert\mathbf{ab}\rvert=\delta\sin(3\alpha/2-45°)/(\sin(45°+\alpha/2)\sin\alpha)$
BK	$L/(2\cos\alpha)$	$\lvert\mathbf{bd}\rvert=\delta\sin(2\alpha)\cos(45°+\alpha/2)/(\sin(45°+\alpha/2)\sin\alpha)$
BN	$L/(2\cos\alpha)$	$\lvert\mathbf{de}\rvert=\delta\sin(3\alpha/2-45°)/(\sin(45°+\alpha/2)\sin\alpha)$
EK	$L/(2\sin(45°+\alpha/2))$	$\lvert\mathbf{ob}\rvert=\delta\sin(2\alpha)/(\sin(45°+\alpha/2)\sin\alpha)$
KN	$L/(2\sin(45°+\alpha/2))$	$\lvert\mathbf{od}\rvert=\delta\sin(2\alpha)/(\sin(45°+\alpha/2)\sin\alpha)$
NC	$L/(2\cos\alpha)$	$\lvert\mathbf{oe}\rvert=\delta/\sin\alpha$

Total dissipation rate:

$$D=2cL\delta\left[\frac{\sin(45°+\alpha/2)+\sin(3\alpha/2-45°)+\sin(2\alpha)\cos(45°+\alpha/2)}{\sin(2\alpha)\sin(45°+\alpha/2)}+\frac{\cos\alpha}{\sin^2(45°+\alpha/2)}\right]$$

Rate of external work:*

$$W=Q_f\,L\delta$$

Upper bound theorem:

$$Q_f\le\min_{\alpha}\left\{2c\left[\frac{\sin(45°+\alpha/2)+\sin(3\alpha/2-45°)+\sin(2\alpha)\cos(45°+\alpha/2)}{\sin(2\alpha)\sin(45°+\alpha/2)}+\frac{\cos\alpha}{\sin^2(45°+\alpha/2)}\right]\right\}$$

$$Q_f\le 5.29\,c\ (\textit{for}\ \alpha=48.6°)$$

*Note: rate of work of body forces equals zero

Figure 6.8 Upper bound solution for the mechanism in Figure 6.5c

of the failure load is $Q_f\le 5.66c$ for $\alpha=54.7°$, i.e. it is actually the same as that based on the previous mechanism.

The last two cases considered in Figure 6.5 represent more elaborated schemes of the former kinematic mechanisms. Both lead, once again, to an identical assessment of the bearing capacity of $Q_f\le 5.29c$, which is lower than that before. The procedure is conceptually the same as that followed in Figures 6.6–6.7 and the details regarding the

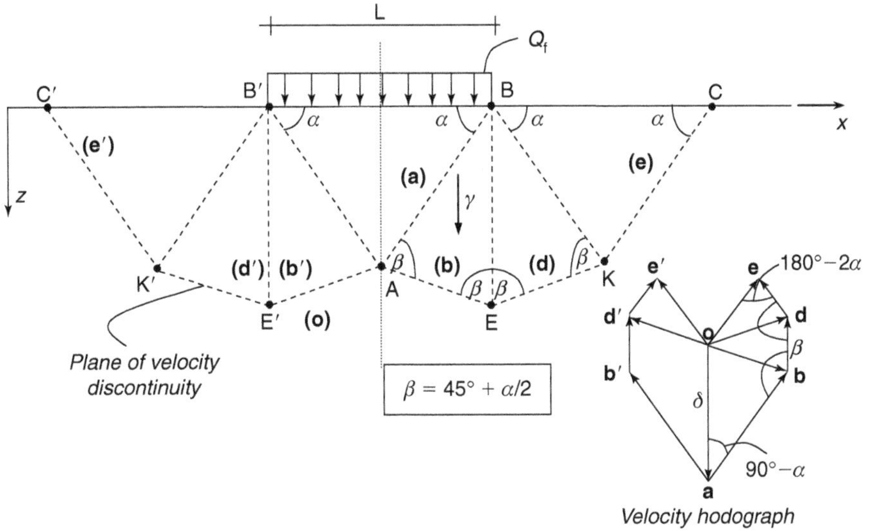

Discontinuity plane	*Length*	*Velocity discontinuity*
BA, B′A	$L/(2\cos\alpha)$	$\lvert\mathbf{ab}\rvert = \lvert\mathbf{ab'}\rvert = \delta$
BE, B′E′	$L/(2\cos\alpha)$	$\lvert\mathbf{bd}\rvert = \lvert\mathbf{b'd'}\rvert = 2\delta\cos\alpha/\tan(45^\circ+\alpha/2)$
AE, AE′	$L/(2\sin(45^\circ+\alpha/2))$	$\lvert\mathbf{ob}\rvert = \lvert\mathbf{ob'}\rvert = \delta\cos\alpha/\sin(45^\circ+\alpha/2)$
BK, B′K′	$L/(2\cos\alpha)$	$\lvert\mathbf{de}\rvert = \lvert\mathbf{d'e'}\rvert = \delta\sin(3\alpha/2-45^\circ)/(2\sin\alpha\sin(45^\circ+\alpha/2))$
EK, E′K′	$L/(2\sin(45^\circ+\alpha/2))$	$\lvert\mathbf{od}\rvert = \lvert\mathbf{od'}\rvert = \delta\cos\alpha/\sin(45^\circ+\alpha/2)$
CK, C′K′	$L/(2\cos\alpha)$	$\lvert\mathbf{oe}\rvert = \lvert\mathbf{oe'}\rvert = \delta/(2\sin\alpha)$

Total dissipation rate:

$$D = cL\delta\left[\frac{1}{\cos\alpha} + \frac{2}{\tan(45^\circ+\alpha/2)} + \frac{\sin(3\alpha/2-45^\circ)}{\sin(2\alpha)\sin(45^\circ+\alpha/2)} + \frac{1}{\sin(2\alpha)} + \frac{2\cos\alpha}{\sin^2(45^\circ+\alpha/2)}\right]$$

Rate of external work:*

$$W = Q_f L\delta$$

Upper bound theorem:

$$Q_f \le \min_{\alpha}\left\{c\left[\frac{1}{\cos\alpha} + \frac{2}{\tan(45^\circ+\alpha/2)} + \frac{\sin(3\alpha/2-45^\circ)}{\sin(2\alpha)\sin(45^\circ+\alpha/2)} + \frac{1}{\sin(2\alpha)} + \frac{2\cos\alpha}{\sin^2(45^\circ+\alpha/2)}\right]\right\}$$

$$Q_f \le 5.29\,c \;(\textit{for } \alpha = 48.6^\circ)$$

*Note: rate of work of body forces equals zero

Figure 6.9 Upper bound solution for the mechanism in Figure 6.5d

specification of the rate of energy dissipation along all discontinuity planes are given in Figures 6.8 and 6.9. Thus, recalling again the lower bound approximation (Figure 6.3), the short-term bearing capacity is now predicted to be within a relatively narrow range of $5c \le Q_f \le 5.29c$.

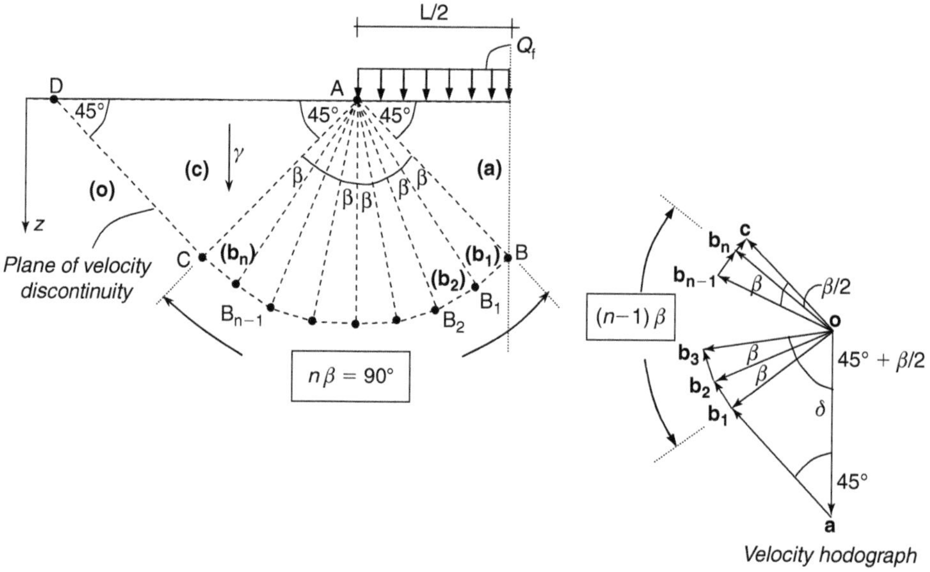

Discontinuity plane	*Length*	*Velocity discontinuity*
AB	$L/\sqrt{2}$	$\lvert ab_1 \rvert = \delta \sin(45° + \beta/2)/\cos(\beta/2)$
$AB_1, AB_2, \ldots, AB_{n-1}$	$L/\sqrt{2}$	$\lvert b_1 b_2 \rvert = \ldots = \lvert b_{n-1} b_n \rvert = \sqrt{2}\delta \tan(\beta/2)$
AC	$L/\sqrt{2}$	$\lvert b_n c \rvert = \delta \tan(\beta/2)/\sqrt{2}$
$BB_1, B_1B_2, \ldots, B_{n-1}C$	$\sqrt{2}L \sin(\beta/2)$	$\lvert ob_1 \rvert = \ldots = \lvert ob_n \rvert = \delta/(\sqrt{2}\cos(\beta/2))$
CD	$L/\sqrt{2}$	$\lvert oc \rvert = \delta/\sqrt{2}$

Total dissipation rate:

$$D = cL\delta[1 + 2n \tan(\beta/2)] = cL\delta[1 + 2n \tan(45°/n)]$$

Rate of external work:*

$$W = Q_f L\delta/2$$

Upper bound theorem:

$$Q_f \le \min_n c[2 + 4n\tan(45°/n)] = c\lim_{n\to\infty}[2 + 4n\tan(45°/n)] = c\left[2 + 4n\frac{\pi}{4n}\right] = c[2+\pi]$$

$$Q_f \le (2+\pi)c$$

*Note: rate of work of body forces equals zero

Figure 6.10 Upper bound solution for an enhanced mechanism with increased number of wedges

The mechanisms identified in Figures 6.7 and 6.9 can be further enhanced by increasing the number of wedges in the region confined between (a) and (c)/(c′) in Figure 6.7, or (a) and (e)/(e′) in Figure 6.9. Similarly as before, the failure zone is said to be symmetric with respect to the centerline and the wedge (a), directly below the foundation, moves downward with the velocity δ (Figure 6.10). The whole mechanism

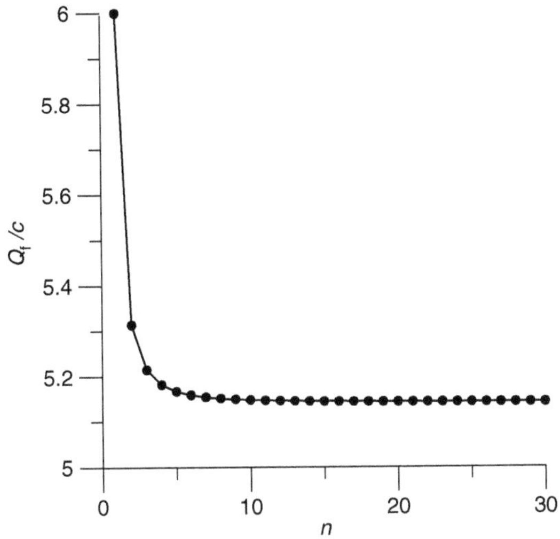

Figure 6.11 Evolution of failure load with number of wedges (n) for the mechanism in Figure 6.10

employs the wedges (**a**) and (**c**), together with a set of n isosceles triangles ($\mathbf{b}_1$)-($\mathbf{b}_n$) in the region ABC. In general, by increasing n, the estimated value of bearing capacity decreases, as evidenced in Figure 6.11. In the limiting case of $n \to \infty$, the minimum value of failure load becomes $Q_f = (2+\pi)c$. This particular value represents, in fact, an *exact* solution that was obtained by Prandtl based on the method of characteristics [50]. The kinematic mechanism associated with this solution is shown in Figure 6.12. Note that in this mechanism, the energy is dissipated not only along the discontinuity planes but also inside the region ABC, in which the shear strain develops. Again, the details of the solution are provided in the table that is appended to Figure 6.12. By comparing the velocity fields of Figures 6.10 and 6.12, it is evident that the kinematic mechanism depicted in Figure 6.10, i.e. the one consisting of a set of n isosceles triangles, is a discrete approximation to the continuum representation in Figure 6.12. The equivalence between the two mechanisms is attained at $n \to \infty$.

(ii) Mohr-Coulomb material

Consider now the case when the conditions at failure are defined in terms of Mohr-Coulomb criterion, i.e.

$$F = |\sigma_1 - \sigma_3| + (\sigma_1 + \sigma_3) - 2c\cos\phi = 0 \tag{6.27}$$

where ϕ is the angle of friction and c is the cohesion.

The lower bound approach requires specification of a statically and plastically admissible stress field, i.e. one that satisfies equilibrium, boundary conditions and does not violate (6.27). In order to illustrate the methodology, consider a simple example that

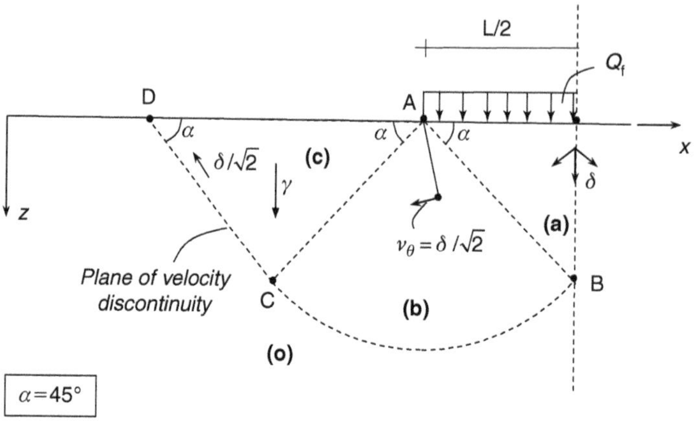

Discontinuity plane	Length	Velocity discontinuity	Dissipation rate
AB	$L/\sqrt{2}$	$\delta/\sqrt{2}$	$cL\delta/2$
BC	$\pi L/(2\sqrt{2})$	$\delta/\sqrt{2}$	$c\pi L\delta/4$
AC	$L/\sqrt{2}$	0	0
CD	$L/\sqrt{2}$	$\delta/\sqrt{2}$	$cL\delta/2$

Rate of disspation along discontinuty planes:

$$D_1 = cL\delta(1+\pi/4)$$

Shear zone	*Strain rates (plane strain case)*	*Dissipation rate* (per unit volume)*
ABC	$\dot{\gamma}_{r\theta} = -\frac{v_\theta}{r} = -\frac{\delta}{\sqrt{2}\,r}, \quad \dot{\varepsilon}_{rr} = 0, \quad \dot{\varepsilon}_{\theta\theta} = 0$	$c\lvert\dot{\gamma}_{r\theta}\rvert$

Rate of disspation within the shear zone:

$$D_2 = \int_0^{L/\sqrt{2}} c\left(\frac{\delta}{\sqrt{2}\,r}\right)\left(\frac{\pi}{2} r\,dr\right) = cL\delta\frac{\pi}{4}$$

Total dissipation rate:

$$D = D_1 + D_2 = cL\delta(1+\pi/2)$$

Rate of external work:

$$W = Q_f\,\delta L/2$$

Upper bound theorem:

$$W \le D \Rightarrow \quad Q_f \le c(2+\pi) \approx 5.14c$$

*see Ref. [49] for a proof

Figure 6.12 Upper bound solution based on Prandtl's mechanism

involves assessment of the critical height of a vertical embankment in a Mohr-Coulomb material. The solution is detailed in Figure 6.13. In the region $0 \le z < H$, a uniaxial stress field is stipulated $\sigma_z = -\gamma z$, $\sigma_x = \sigma_y = 0$ that is said to satisfy (6.27) as $z \to H$. In the remaining sub-regions a lateral stress is added so that $F < 0$. The condition of

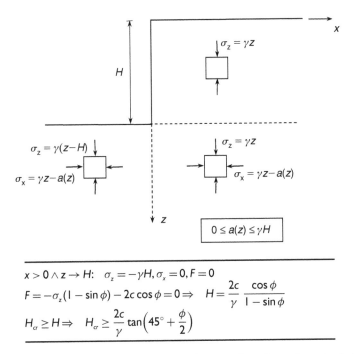

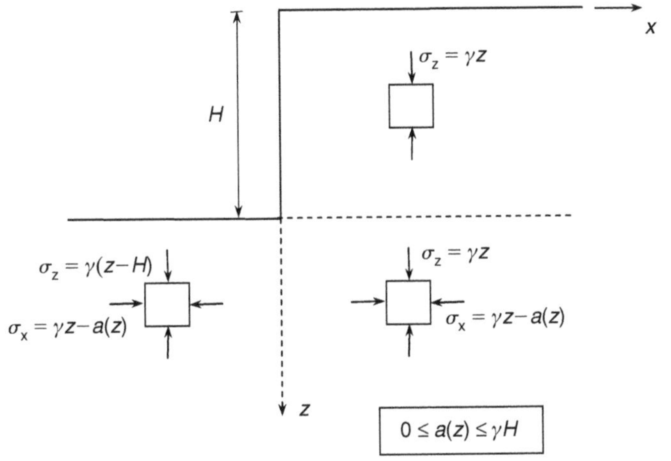

$$x > 0 \wedge z \to H: \quad \sigma_z = -\gamma H, \sigma_x = 0, F = 0$$

$$F = -\sigma_z(1 - \sin\phi) - 2c\cos\phi = 0 \Rightarrow \quad H = \frac{2c}{\gamma}\frac{\cos\phi}{1 - \sin\phi}$$

$$H_{cr} \ge H \Rightarrow \quad H_{cr} \ge \frac{2c}{\gamma}\tan\left(45^\circ + \frac{\phi}{2}\right)$$

Figure 6.13 Assessment of critical height of a vertical cut; lower bound solution

$F = 0$ at $z \to H$ provides an estimate on H as a function of ϕ, c, γ, and by the virtue of the lower bound theorem there is $H_{cr} \ge H$.

A simple upper bound solution is given in Figure 6.14. Here, a plane of velocity discontinuity CB is introduced that originates at the toe of the embankment. The solution procedure is analogous to that outlined earlier and the details are again provided in the table attached to the figure. In this case, the velocity discontinuity vector deviates from the direction of the discontinuity plane by the angle ϕ. The rate of energy dissipated along CB is defined by representation (6.25) and the external work is done by the gravity forces. For this particular mechanism, the upper bound estimate differs from the lower bound by a factor of 2. Note that a more accurate solution can be obtained by assuming that the discontinuity plane is a logarithmic spiral (see Drucker and Prager [51]).

By examining the solutions given in Figures 6.13 and 6.14, it is evident that the stability of the vertical cut cannot be maintained if the material is cohesionless, i.e. $c = 0$. In this case, a retaining support is required to ensure a safe configuration. A common solution is to construct a vertical sheet pile wall, as indicated in Figure 6.15. In order to enforce the stability of the embankment, the wall must penetrate into the soil. Thus, the problem here is to assess the minimum depth of penetration h that is required.

An upper bound assessment can be obtained by employing a kinematically admissible velocity field. Figure 6.15 shows a simple mechanism that incorporates two rigid wedges (**a**) and (**b**). For a cohesionless material, the rate of energy dissipation along

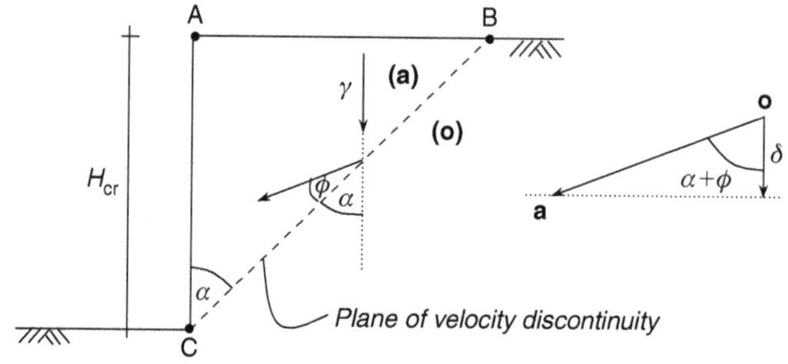

Rate of dissipation:

$$D = c|\mathbf{oa}|\cos\phi|\,BC| = c\,\delta H_{cr}\frac{\cos\phi}{\cos\alpha\cos(\alpha+\phi)}$$

Rate of external work:

$$W = \frac{1}{2}\gamma H_{cr}|\,AB|\delta = \frac{1}{2}\gamma\,\delta\,H_{cr}^2\tan\alpha$$

Upper bound theorem:

$$W \le D \Rightarrow H_{cr} \le \min_{\alpha\in(0^\circ,90^\circ)}\frac{2c}{\gamma}\left\{\frac{\cos\phi}{\sin\alpha\cos(\alpha+\phi)}\right\} = \frac{4c}{\gamma}\tan\left(45^\circ+\frac{\phi}{2}\right)\left(\mathit{for}\ \alpha = 45^\circ - \frac{\phi}{2}\right)$$

Figure 6.14 Assessment of critical height of a vertical cut; upper bound solution

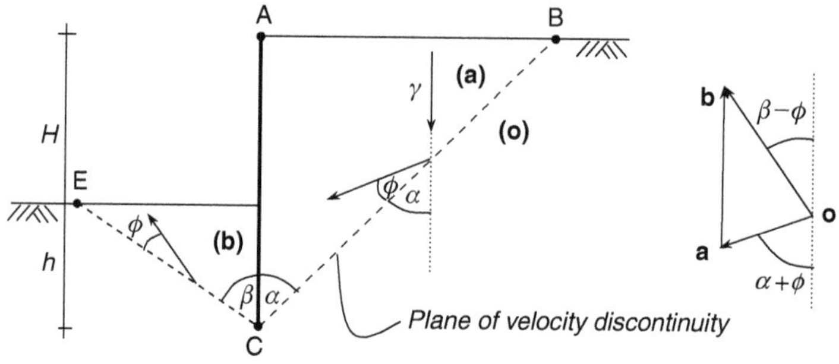

Figure 6.15 Assessment of minimum depth of penetration of a vertical sheet pile wall; upper bound solution for a cohesionless soil

the discontinuity planes is equal to zero, eq.(6.25). Thus, the statement of the upper bound theorem reads $W \le 0$, where W is the rate of work of body forces, i.e.

$$W = \frac{1}{2}\gamma|\mathbf{oa}|\cos(\alpha+\phi)\left[(H+h)^2\tan\alpha - h^2\tan\beta\frac{\tan(\alpha+\phi)}{\tan(\beta-\phi)}\right] \tag{6.28}$$

In this case, the problem to be solved can be phrased as follows: given the parameters H, γ, ϕ determine the value of h for which $W \le 0$ for all physically admissible values of α and β. By examining the velocity hodograph of Figure 6.15, it is noted that for $\beta-\phi<0$ the wedges (a) and (b) will move sideways and upwards. Similarly, for $\alpha+\phi>90°$, the vertical motion of the wedges will be upwards. Thus, in both these cases the rate of external work will become negative. Thus, the constraint $W \leq 0$ requires finding the value of h such that

$$g(h)=(H+h)^2\tan\alpha-h^2\tan\beta\frac{\tan(\alpha+\phi)}{\tan(\beta-\phi)}\leq 0 \tag{6.29}$$

for all α and β satisfying $0<\alpha\leq(90°-\phi)$, $\phi<\beta\leq 90°$.
Given the constraints on α and β, the following trigonometric inequalities hold

$$\tan\alpha<\tan(\alpha+\phi), \tan(\beta-\phi)<\tan\beta \Rightarrow \tan\alpha-\tan\beta\frac{\tan(\alpha+\phi)}{\tan(\beta-\phi)}<0 \tag{6.30}$$

Thus, $g(h)$ is a quadratic function

$$g(h)=\left(\tan\alpha-\tan\beta\frac{\tan(\alpha+\phi)}{\tan(\beta-\phi)}\right)h^2+(2H\tan\alpha)h+H^2\tan\alpha \tag{6.31}$$

in which the coefficient at h^2 is negative.
The quadratic form (6.31) has two real roots $h_1<h_2$, so that in order to enforce the inequality (6.29) there must be $h>h_2$, i.e.

$$h\geq \max_{\substack{\alpha\in(0,90°-\phi]\\ \beta\in(\phi,90°]}} H\left\{1/\left(\sqrt{\frac{\tan\beta}{\tan(\beta-\phi)}\frac{\tan(\alpha+\phi)}{\tan\alpha}}-1\right)\right\} \tag{6.32}$$

The maximum is attained here for

$$\begin{aligned}\min_{\beta\in[\phi,90°)}\left\{\frac{\tan\beta}{\tan(\beta-\phi)}\right\}&=\frac{\tan(45°+\phi/2)}{\tan(45°-\phi/2)}\\ \min_{\alpha\in(0,90°-\phi]}\left\{\frac{\tan(\alpha+\phi)}{\tan\alpha}\right\}&=\frac{\tan(45°+\phi/2)}{\tan(45°-\phi/2)}\end{aligned} \tag{6.33}$$

which yields, after some algebraic transformations

$$h\geq\frac{H}{\tan^2(45°+\phi/2)-1} \tag{6.34}$$

The above expression represents an upper bound estimate for the depth of penetration of the sheet pile wall that is required to maintain the stability of the embankment.

In conclusion, it should be mentioned again that the assessment of collapse load based on limit analysis is not affected by the elastic properties of the material. This is due to the fact that at the impending collapse, or the onset of unlimited plastic flow, the stress rates vanish (cf. Ref.[44]). Thus, since $\dot{\sigma}_{ij}\to 0\Rightarrow\dot{\varepsilon}_{ij}\to\dot{\varepsilon}^p_{ij}$, the body behaves as though it were rigid-plastic. Hence, all lower/upper bound solutions presented here, for both Tresca and Mohr-Coulomb material, are valid for elastic – plastic idealization as well.

Chapter 7

Description of inherent anisotropy in geomaterials

The presentation so far has been restricted to materials which are isotropic at the macroscale. However, many geomaterials display inherent anisotropy that is intrinsically related to their microstructure. Examples include sedimentary rocks/soils that are typically formed by deposition and progressive consolidation during diagenesis. Such formations are characterized by the appearance of multiple sedimentary layers which can often be identified by a visual examination. In granular media, the preferred orientation of microstructure can also be associated with a parallel alignment of particles, which is often encountered in river, beach and dune sand.

The primary manifestation of anisotropy associated with existence of a distinct microstructure is a strong directional dependence of both strength and deformation characteristics. The description of these properties employs, in general, two distinct mathematical approaches, i.e. a macro and micromechanical one. In the former case, which is of primary interest here, the material is conceptually considered as a continuum; the approach is strictly phenomenological while the anisotropy effects are described by incorporating some tensorial measures of material microstructure. On the other hand, the micromechanical approach consists, in essence, of specifying the average mechanical response in terms of properties of constituents (which are usually isotropic) and the actual geometry of the microstructure. Such an approach is commonly referred to as *homogenization*.

The use of anisotropic materials is often unavoidable for many geotechnical works associated with foundations, excavations, tunneling, etc. Therefore, an appropriate consideration of directional effects is of importance in the context of both analysis and design. In this chapter, the issues of the specification of conditions at failure and the description of deformation process in inherently anisotropic geomaterials are addressed. The focus is on a macroscopic approach. In particular, two distinct formulations are outlined (after Refs.[52,53]); one referred to as a critical plane approach and the other incorporating a scalar anisotropy parameter that is a function of mixed invariants of stress and microstructure-orientation tensors.

7.1 FORMULATION OF ANISOTROPIC FAILURE CRITERIA

The description of mechanical response of inherently anisotropic media requires the specification of conditions at failure under an arbitrary stress state. Over the last few decades, numerous failure criteria had been proposed; a review of different methodologies can be found, for example, in Ref. [54]. One of the approaches to define the failure function is to invoke linear as well as higher order terms in stress components referred to the coordinate system associated with the axes of symmetry of the material. An example of such an approach is an extension of the well-known Hill's criterion [55], as proposed by Tsai and Wu [56]. The latter employs a polynomial representation that retains only linear and second order terms in stress components. A more rigorous approach, which makes use of general representation theorems and employs ten independent basic and mixed invariants of stress and microstructure tensors, was proposed by Boehler and Sawczuk [57,58]. Using a conceptually similar approach, Cowin [59] developed a simplified approximation, in which the failure criterion is defined as a quadratic function of stress and fabric tensors. The primary difficulty with implementation of these approaches is the fact that they require a large number of material functions/parameters to be identified. For example, the simplified approximation developed by Cowin [59] employs 12 independent functions of material microstructure, the specification of which is rather ambiguous.

In this section, some simplified approaches are discussed which employ the methodology outlined by Pietruszczak and Mroz in Refs. [52,53]. The critical plane approach, which incorporates a spatial distribution of strength parameters, is described first. Later, an alternative framework is presented that employs scalar anisotropy parameters which depend on the orientation of the principal stress triad relative to the preferred directions of microstructure.

7.1.1 Specification of failure criteria based on critical plane approach

In this approach, the failure criterion is defined in terms of traction components acting on the critical/localization plane. The approach employs a spatial distribution of strength parameters and the orientation of the critical plane is determined by maximizing the failure function with respect to the orientation.

In order to illustrate the approach, consider a failure function that incorporates a single strength parameter c, i.e.

$$F = f(t_i^n, t_i^s) - c(n_i) \tag{7.1}$$

Here, t_i^n, t_i^s are the normal and tangential components of traction vector t_i acting on a plane with unit normal n_i, so that

$$t_k^n = \sigma_{ij} n_i n_j n_k; \quad t_k^s = (\delta_{ki} - n_k n_i)\sigma_{ij} n_j \tag{7.2}$$

The failure on this plane is said to commence when $F = 0$, i.e. $f = c$, while the strength parameter c is assumed to be orientation-dependent, i.e. $c = c(n_i)$. The function $c(n_i)$ is a scalar valued function defined over a unit sphere and its representation may be assumed in the form analogous to that employed in Refs. [52,60], i.e.

$$c(n_i) = c_0(1 + \Omega_{ij} n_i n_j + \Omega_{ijkl} n_i n_j n_k n_l + \Omega_{ijklmn} n_i n_j n_k n_l n_m n_n + \cdots) \tag{7.3}$$

In the above equation, c_0 is a constant and Ω's are symmetric *traceless* tensors of even rank describing the bias in the spatial distribution of $c(n_i)$. Assuming, for simplicity, that $\Omega_{ijkl} = b_1 \Omega_{ij} \Omega_{kl}$ and $\Omega_{ijklmn} = b_2 \Omega_{ij} \Omega_{kl} \Omega_{mn}$, ..., yields

$$c(n_i) = c_0(1 + \Omega_{ij} n_i n_j + b_1 (\Omega_{ij} n_i n_j)^2 + b_2 (\Omega_{ij} n_i n_j)^3 + \cdots) \tag{7.4}$$

The representation (7.4) employs only a second-order tensor Ω_{ij} whose eigenvectors coincide with the principal material axes. For an orthotropic material there are two independent eigenvalues of Ω_{ij} (since $\Omega_{ii} = 0$). The number reduces further to one for a transversely isotropic material, while for an isotropic material Ω_{ij} vanishes. Note that if the higher order terms in (7.4) are neglected, i.e. $b_1 = b_2 = \ldots\ldots = 0$, the constant c_0 is the orientation average of $c(n_i)$.

The mechanical behaviour within the plane is assumed to be isotropic, so that f is taken as an isotropic function of t_i, i.e. $f(t_i^n, t_i^s) = f(\sigma, \tau)$ where σ, τ are defined, according to eq.(7.2), as $\sigma = t_i n_i$, $\tau = |t_i^s|$. Thus, the problem of onset of failure and specification of the orientation of the critical plane can be formulated as a constrained optimization problem, i.e.

$$\max_{n_i} F = \max_{n_i} (f(\sigma, \tau) - c(n_i)) = 0; \quad n_i n_i = 1 \tag{7.5}$$

Alternatively, one can define the tangential traction as $\tau = |\sigma_{ij} n_j s_i|$ where s_i is an arbitrary unit vector normal to n_i. In this case, the problem can also be formulated as

$$\max_{n_i, s_i} F = \max_{n_i, s_i} (f(\sigma, \tau) - c(n_i)) = 0; \quad n_i n_i = 1, \; n_i s_i = 0, \; s_i s_i = 1 \tag{7.6}$$

Equations (7.5) and/or (7.6) can be solved by Lagrange multipliers or any other suitable optimization technique. The solution provides the orientation of the critical/localization plane and defines the conditions at which the failure occurs. Examples of different criteria are provided below.

(i) Tresca criterion

Consider first Tresca's hypothesis according to which the failure takes place if the shear stress τ on the localization plane reaches a critical value c. Assuming, for simplicity, that $b_1 = b_2 = \ldots\ldots = 0$ in eq.(7.4), so that $c(n_i)$ is linear in $\Omega_{ij} n_i n_j$, the failure criterion (7.6) may be written as

$$\max_{n_i, s_i} F = \max_{n_i, s_i} (\tau - c(n_i)) = 0; \quad c = c_0(1 + \Omega_{ij} n_i n_j) \tag{7.7}$$

In this case, the failure function takes the form

$$F = |\sigma_{ij} n_i s_j| - c_0(1 + \Omega_{ij} n_i n_j) \tag{7.8}$$

and the Lagrangian function, accounting for the constraints given in (7.6), becomes

$$G = |\sigma_{ij} n_i s_j| - c_0(1 + \Omega_{ij} n_i n_j) - \lambda_1(n_i n_i - 1) - \lambda_2 n_i s_i - \lambda_3(s_i s_i - 1) \tag{7.9}$$

where λ_1, λ_2 and λ_3 are the Lagrange multipliers. The stationary conditions of G with respect to n_i, s_i yield a set of equations

$$\begin{aligned}\frac{\partial G}{\partial n_i} &= (S\sigma_{ij} - \lambda_2 \delta_{ij})s_j - 2(c_0 \Omega_{ij} + \lambda_1 \delta_{ij})n_j = 0 \\ \frac{\partial G}{\partial s_i} &= (S\sigma_{ij} - \lambda_2 \delta_{ij})n_j - 2\lambda_3 s_i = 0\end{aligned} \tag{7.10}$$

where $S = \mathrm{sgn}(\sigma_{ij} n_i s_j)$. At the same time, the stationary conditions with respect to λ's provide the constraints of the problem. The resulting algebraic equations can be solved for both the direction cosines n_i and s_i, as well as the Lagrange multipliers. Given n_i and s_i, the value of failure function (7.8) can then be evaluated to determine whether the conditions at failure have been reached.

The problem can also be formulated in terms of n_i alone, i.e. viz. eq.(7.5). In view of eq.(7.2), there is

$$t_i^s t_i^s = \tau^2 = (\delta_{jk} - n_j n_k) t_j t_k; \quad t_i = \sigma_{ij} n_j \tag{7.11}$$

so that the Lagrangian function G may be written in the form

$$G = t_i^s t_i^s - c_0^2 (1 + \Omega_{ij} n_i n_j)(1 + \Omega_{kl} n_k n_l) - \lambda (n_i n_i - 1) \tag{7.12}$$

Thus, substituting eq.(7.11) into (7.12) and differentiating, one obtains

$$\frac{\partial G}{\partial n_i} = (\sigma_{kj}\sigma_{ki} - 2c_0^2 \Omega_{ij}) n_j - 2(\sigma_{ij}\sigma_{kl} + c_0^2 \Omega_{ij}\Omega_{kl}) n_j n_k n_l - \lambda n_i = 0 \tag{7.13}$$

Again, the above set of three scalar equations, together with the constraint $n_i n_i = 1$, can be solved for the direction cosines n_i and the multiplier λ. Note that for an isotropic material, i.e. when $\Omega_{ij} = 0$, the problem reduces to finding an orientation which maximizes the shear stress τ. In this case, the Lagrangian function G in eq.(7.12) will reach the maximum on the plane that bisects the right angle between the directions of the major and minor principal stresses.

(ii) Tension cut-off criterion

This is relevant to a class of materials that have the ability to resist tension. The simplest criterion is based on the notion that tensile failure/fracture occurs when the normal component of the traction vector t_i acting on the localization plane reaches a critical value c. Thus, the strength parameter $c(n_i)$, viz. eq.(7.4), is now identified with the maximum admissible value of $\sigma = t_i n_i$. Assuming again that $c(n_i)$ is linear in $\Omega_{ij} n_i n_j$, the failure criterion may be written as

$$\max_{n_i} F = \max_{n_i} (\sigma - c(n_i)) = 0; \quad c = c_0 (1 + \Omega_{ij} n_i n_j) \tag{7.14}$$

In this case, the failure function is defined as

$$F = \sigma_{ij} n_i n_j - c_0 (1 + \Omega_{ij} n_i n_j) \tag{7.15}$$

while the Lagrangian function becomes

$$G = \sigma_{ij}n_i n_j - c_0(1 + \Omega_{ij}n_i n_j) - \lambda(n_i n_i - 1) \tag{7.16}$$

The stationary conditions with respect to n_i take the form

$$\frac{\partial G}{\partial n_i} = 2(\sigma_{ij}n_j - c_0\Omega_{ij}n_j) - 2\lambda\delta_{ij}n_j = 0 \tag{7.17}$$

from which

$$(B_{ij} - \lambda\delta_{ij})n_j = 0; \quad B_{ij} = \sigma_{ij} - c_0\Omega_{ij} \tag{7.18}$$

Eq.(7.18) defines an eigenvalue problem that can be solved to specify the direction cosines n_i. Clearly, if $\Omega_{ij} = 0$ then the unit normal to the critical plane will be coaxial with the direction of the maximum tensile stress. Apparently, the representation (7.14), as well as that corresponding to Tresca criterion (7.7), can be augmented by introducing higher order terms in $c(n_i)$, eq.(7.4), if the experimental evidence warrants it.

(iii) Coulomb criterion

Assuming that both frictional and cohesive properties are orientation-dependent, and taking again a linear form of (7.4), i.e. $b_1 = b_2 = \ldots\ldots = 0$, this criterion can written as

$$\max_{n_i,\, s_i} F = \max_{n_i,\, s_i}(\tau + \mu\sigma - c) = 0; \quad \mu = \mu_0(1+\Omega^{\mu}_{kl}n_k n_l), \quad c = c_0(1+\Omega^{c}_{ij}n_i n_j) \tag{7.19}$$

where μ is identified with the tangent of internal friction angle and c refers now to cohesion. According to (7.19), the failure function F can be expressed as

$$F = |\sigma_{ij}n_i s_j| + \mu_0(1 + \Omega^{\mu}_{kl}n_k n_l)\sigma_{ij}n_i n_j - c_0(1 + \Omega^{c}_{ij}n_i n_j) \tag{7.20}$$

and the Lagrangian function G can be constructed as follows

$$\begin{aligned} G = {} & |\sigma_{ij}n_i s_j| + \mu_0(1 + \Omega^{\mu}_{kl}n_k n_l)\sigma_{ij}n_i n_j - c_0(1 + \Omega^{c}_{ij}n_i n_j) - \lambda_1(n_i n_i - 1) \\ & -\lambda_2 n_i s_i - \lambda_3(s_i s_i - 1) \end{aligned} \tag{7.21}$$

The stationary conditions of G with respect to n_i, s_i become

$$\begin{aligned} \frac{\partial G}{\partial n_i} &= (S\sigma_{ij} - \lambda_2\delta_{ij})s_j - 2(-\mu_0\sigma_{ij} + c_0\Omega^{c}_{ij} + \lambda_1\delta_{ij})n_j \\ &\quad +2\mu_0(\Omega^{\mu}_{ip}\sigma_{jk} + \Omega^{\mu}_{jp}\sigma_{ik})n_p n_j n_k = 0 \\ \frac{\partial G}{\partial s_i} &= (S\sigma_{ij} - \lambda_2\delta_{ij})n_j - 2\lambda_3 s_i = 0 \end{aligned} \tag{7.22}$$

whereas the stationary conditions with respect to λ's provide, once more, the constraints of the problem, i.e $n_i n_i = 1$, $n_i s_i = 0$, $s_i s_i = 1$. It should be noted that for a cohesionless material there is $c_0 = 0$. At the same time, the formulation corresponding to Tresca criterion can be recovered by setting $\mu_0 = 0$. Again, the set of simultaneous equations (7.22) can be solved to define the orientation of the critical plane as well the Lagrangian multipliers. For an isotropic material, i.e. when $\Omega^{\mu}_{ij} = \Omega^{c}_{ij} = 0$, the normal n_i to the critical plane lies in the plane containing the major and minor principal stresses and its direction deviates from that of the major principal stress axis by $\pi/4 + \phi/2$, where $\phi = const$ is the friction angle (cf. Section 2.4.1). Note that ϕ refers here to frictional properties of isotropic material and cannot be physically identified with $\phi_0 = \tan^{-1}\mu_0$ which defines the orientation average for an anisotropic one.

(iv) Numerical examples

In order to illustrate the approach outlined in this section, consider first a transversely isotropic cohesionless material, for which the onset of failure is governed by Coulomb's criterion (7.19). Since in this case $c_0 = 0$, the specification of conditions at failure requires the identification of one material function $\mu = \mu(n_i)$. The latter can be accomplished by conducting a series of direct shear tests at different orientation of the preferred material axis relative to the direction of shearing. If the higher order terms in the distribution of $\mu(n_i)$ are neglected, i.e. $b_1 = b_2 = \ldots = 0$ in eq.(7.4), the identification entails only the results for samples tested in the directions along and normal to the plane of material symmetry. Refer the problem to the coordinate system indicated in Figure 7.1. Here, $\{x, y, z\}$ is the global frame of reference, while the base vectors $e_i^{(1)}, e_i^{(2)}, e_i^{(3)}$, associated with the coordinate axes $\{x_1, x_2, x_3\}$, define the principal material triad. In this case, $\Omega_1 = \Omega_3$ and since Ω_{ij} is a traceless operator (i.e., $\Omega_1 + \Omega_2 + \Omega_3 = 0$), the following representation is obtained

$$\begin{aligned}\mu(n_i) &= \mu_0(1 + \Omega_{ij}n_i n_j) = \mu_0(1 + \Omega_1 n_1^2 + \Omega_2 n_2^2 + \Omega_3 n_3^2) \\ &= \mu_0(1 + \Omega_1(1 - 3\cos^2\beta))\end{aligned} \tag{7.23}$$

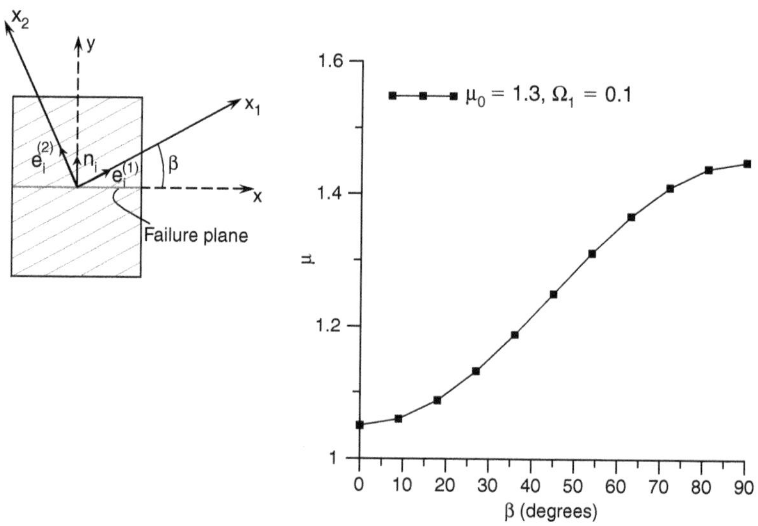

Figure 7.1 Variation of $\mu = \tan\phi$ with orientation of the failure plane (dense sand)

Let the values of μ for samples tested in direct shear at $\beta = 0°$ and $90°$, be $\mu_{\parallel}$ and $\mu_{\perp}$, respectively. Then, according to eq.(7.23)

$$\mu_0 = (\mu_{\parallel} + 2\mu_{\perp})/3; \quad \Omega_1 = \mu_{\perp}/\mu_0 - 1 \tag{7.24}$$

Figure 7.1 shows the distribution of μ as a function of the angle β. The plot corresponds to some typical values representative of a dense sand, i.e.

$$\mu_{\parallel} = 1.05, \ \mu_{\perp} = 1.45 \quad \Rightarrow \quad \mu_0 = 1.3, \ \Omega_1 = 0.1$$

Given the above parameters, the conditions at failure under a general stress state can be investigated by solving the optimization problem (7.22). Consider, for example, the response under drained axial compression. Assume that tests are carried out in *plane strain* conditions under the confinement of σ_0, Figure 7.2, i.e. $\sigma_y = \sigma_1$, $\sigma_x = \sigma_3 = \sigma_0 =$ *const.* and $\sigma_3 < \sigma_z = \sigma_2 < \sigma_1$. The key results are presented in Figure 7.2, which shows the variation of normalized axial strength ($\sigma_1/2\sigma_0$) and the orientation of the critical plane as a function of the orientation of the sample. It is evident that the strength is highest in the vertical sample. For inclined specimens, the strength is progressively

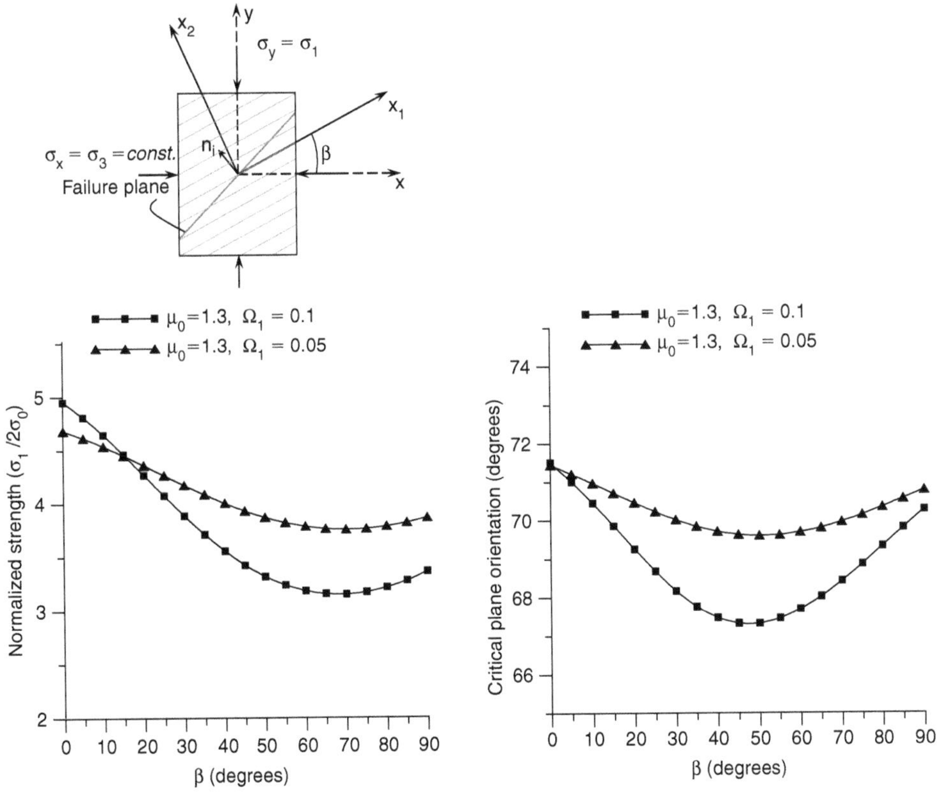

Figure 7.2 Variation of normalized axial strength and orientation of the critical plane with sample orientation; plane strain compression (dense sand)

reduced and reaches a minimum at $\beta \approx 60°$. The orientation of the critical plane also depends on the orientation of the sample and ranges between 67° and 72° with respect to the horizontal axis. The results also show the influence of the degree of anisotropy that is controlled by the eigenvalues of Ω_{ij}. It is clear that a decrease in the value of Ω_1 leads to reduction in the bias of the strength distribution.

The other example given here involves a rock-like material for which the tensile strength is governed by the criterion (7.14). The strength parameter c is now identified with the critical normal stress $c = \max \sigma$ acting on a plane with unit normal n_i. Again, if only a linear term in the distribution of $c(n_i)$, eq.(7.4), is retained, i.e. $b_1 = b_2 = \ldots = 0$, then two independent material constants need to be identified, viz. c_0 and Ω_1. Consider, as an example, schist that is a metamorphic rock having a foliated, or plated, structure. In this case, both material parameters can be specified from direct tension tests that are conducted along the principal material axes, as for tension in the direction of schistosity, or perpendicular to it, the critical plane remains normal to the direction of loading. Denoting the respective strengths as $c_{\parallel}$ and $c_{\perp}$, and employing the representation analogous to (7.23), i.e.

$$c(n_i) = c_0(1 + \Omega_1 n_1^2 + \Omega_2 n_2^2 + \Omega_3 n_3^2) = c_0\left(1 + \Omega_1(1 - 3\cos^2\beta)\right) \tag{7.25}$$

there is

$$c_0 = (c_{\parallel} + 2c_{\perp})/3; \qquad \Omega_1 = c_{\perp}/c_0 - 1 \tag{7.26}$$

Figure 7.3 shows the spatial distribution of critical stress $c = \max \sigma$ along planes at different orientation β with respect to direction of schistosity. Both Cartesian and polar plots are provided that correspond to

$$c_{\parallel} = 2.5\,\text{MPa}, \ c_{\perp} = 8.7\,\text{MPa} \quad \Rightarrow \quad c_0 = 6.6\,\text{MPa}, \ \Omega_1 = 0.3$$

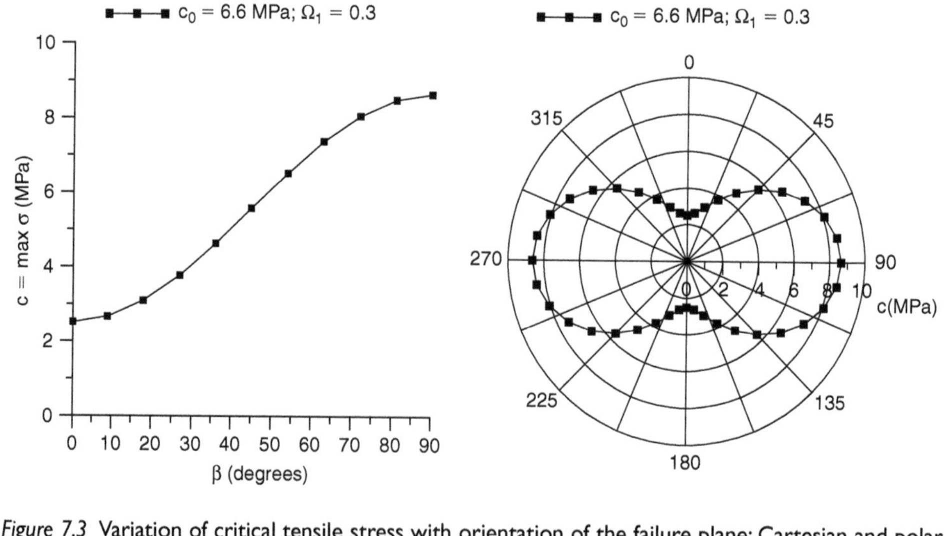

Figure 7.3 Variation of critical tensile stress with orientation of the failure plane; Cartesian and polar representation (schistose rock)

which are the values representative of schistose rock samples tested in Ref.[61].

Given the values of the material constants c_0 and Ω_1, the conditions at failure under a general stress state can be examined. Consider, for example, the response of a sample subjected to axial tension $\sigma_1 = \sigma_t$ along the vertical y-axis at different inclinations of schistosity, Figure 7.4. In this case, the maximization problem (7.5) reduces to a simple form

$$F = \sigma \cos^2 \alpha - c_0 \{1 + \Omega_1 [1 - 3\cos^2(\alpha - \beta)]\} = 0 \tag{7.27}$$

$$\frac{dF}{d\alpha} = -\sigma \sin 2\alpha - 3c_0\Omega_1 \sin 2(\alpha - \beta) = 0$$

which, for a fixed value of β, provides a set of 2 equations for 2 unknowns, i.e. α and σ_t. Here, α defines the orientation of the critical plane, while σ_t is the tensile strength.

The graphs in Figure 7.4 show the distribution of both the strength and orientation of critical plane as a function of inclination of schistosity planes β. For vertical and horizontal samples ($\beta = 0°, 90°$), the strength corresponds to $c_\parallel$ and $c_\perp$, respectively.

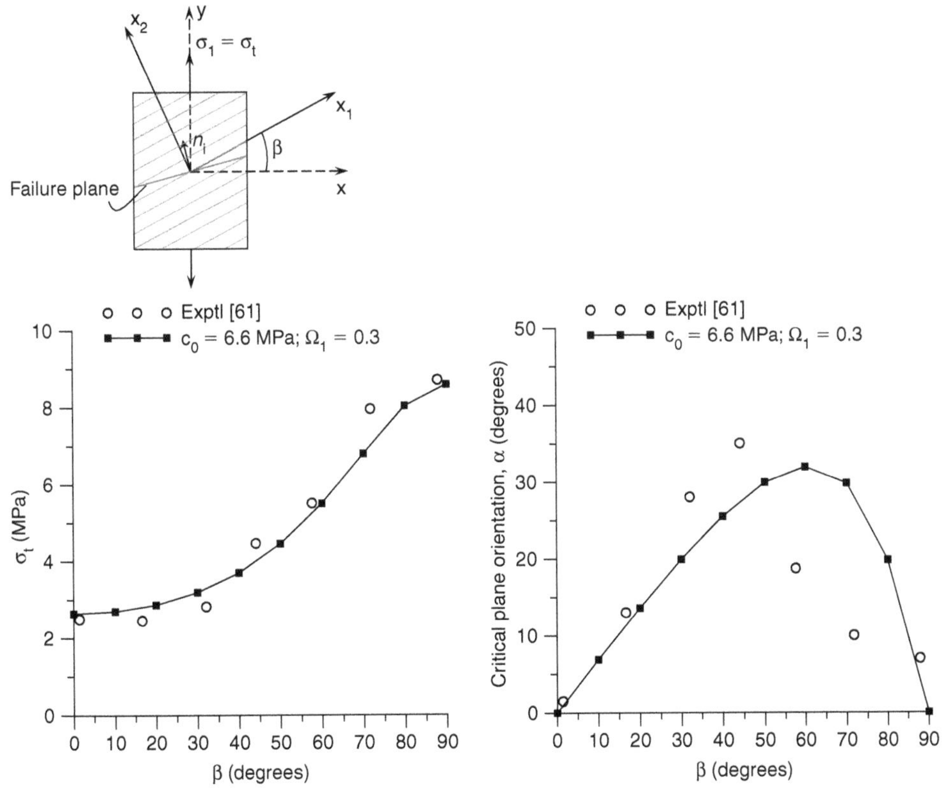

Figure 7.4 Variation of axial tensile strength σ_t and orientation of the critical plane with inclination of schistosity planes β

For inclined specimens, the strength progressively increases reaching a maximum of $\sigma_t = c_\perp$ at $\beta = 90°$. The orientation of the critical plane also depends on the orientation of the sample and ranges between 0° and 35° with respect to the horizontal axis. The results of numerical simulations are compared here with those of the laboratory tests reported in Ref. [61]. The predictions are, in general, consistent with the experimental data; a better quantitative assessment though clearly requires more elaborated forms of the distribution of $c(n_i)$.

The issue of the accuracy of predictions is addressed in more detail in the last example that is provided here. This example incorporates the results of a series of direct shear tests performed on crushed limestone sand with elongated angular-shaped particles, the results of which have been reported in Ref. [62]. The samples were prepared by the sand rain method and were tested at different deposition angles ranging from 0° (horizontally aligned particles) to 90° (vertically aligned). The microstructure was analogous to that depicted in Figure 7.1, with x_2-axis defining the preferred orientation and β being the angle of deposition.

Figure 7.5a shows the best-fit approximations of Coulomb failure envelopes for samples prepared at $\beta = 0°$ and $\beta = 90°$. It is interesting to note that the material develops a residual cohesion, and thus a residual resistance to tension, due to interlocking of particles. A complete set of experimental data, in terms of variation of μ and c with the deposition angle β, is provided in Figures 7.5b and 7.5c. It is evident here that, for this particular material, higher order terms in the representation (7.4) are required to obtain an accurate approximation. In fact, the linear form ($b_1 = b_2 = \ldots = 0$), which corresponds to $\mu_\parallel = 1.05$ and $\mu_\perp = 1.45$ (note that these values are the same as those given in the first example, Figure 7.1), significantly overestimates the friction angle over a wide range of orientations. A much better approximation is obtained by incorporating the dyadic products of degree up to 2 and 4. Note that in this case any commercially available computer algebra system (CAS), i.e. software that facilitates symbolic mathematics (e.g., Mathematica, Matlab, etc.), can be used to obtain the coefficients of the best-fit approximation. In general, the representation including 2nd order term is sufficiently accurate and corresponds to

$$\mu_0 = 1.07, \quad \Omega_1 = 0.23, \quad b_1 = 2.07$$

A similar conclusion is reached in the context of the bias in the spatial variation of cohesion, c, Figure 7.5c. In this case, the second-order approximation is accurate enough and it corresponds to

$$c_0 = 3.51\,\text{kPa}, \; \Omega_1 = 0.22, \; b_1 = 0.82$$

Again, given the parameters above, the conditions at failure under a general stress state can be investigated by solving the constrained optimization problem (7.6). Figure 7.6 show the variation of axial compressive strength with the deposition angle β. The results of simulations correspond to a series of '*triaxial*' *tests* conducted at the confinement of $p_0 = -\sigma_2 = -\sigma_3 = 10\,\text{kPa}$. The experimental data, which is taken again from Ref. [62], is quite restricted as it involves the tests at $\beta = 0°$ and $\beta = 90°$ only. This stems from the fact that for intermediate values of β, the sample has a tendency to distort. In a 'triaxial' test, however, this distortion is constrained by the presence of

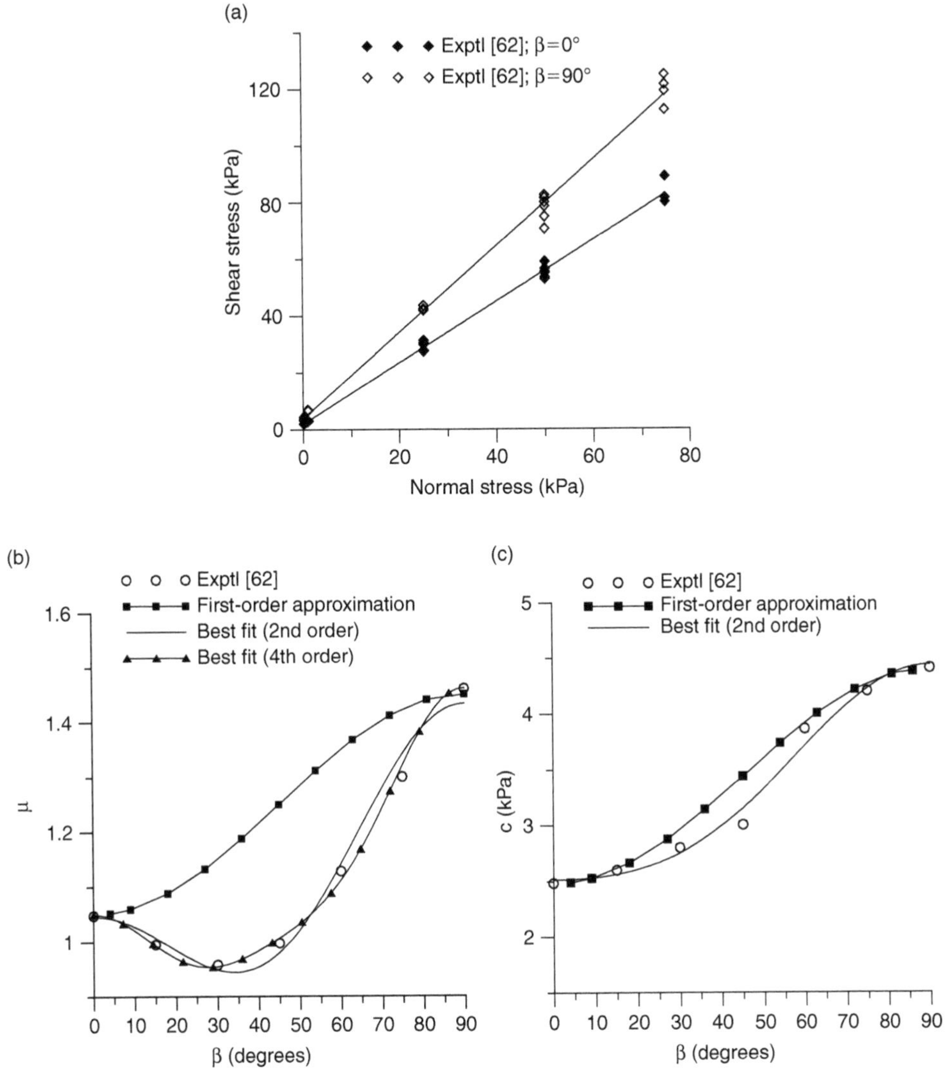

Figure 7.5 Results of direct shear tests on crushed limestone sand [62]; (a) best-fit approximations of failure envelopes, (b,c) approximations to spatial variation of the friction coefficient μ and cohesion c

loading platens. Consequently, the stress field is no longer uniform and the results are not reliable. The numerical simulations given here employed the first and the second-order terms in the distribution of $\mu(n_i)$, respectively, together with the terms of up to degree two in $c(n_i)$. Given the earlier comments, it appears that the linear approximation will again overestimate the axial strength and better estimates will be obtained by incorporating higher order terms.

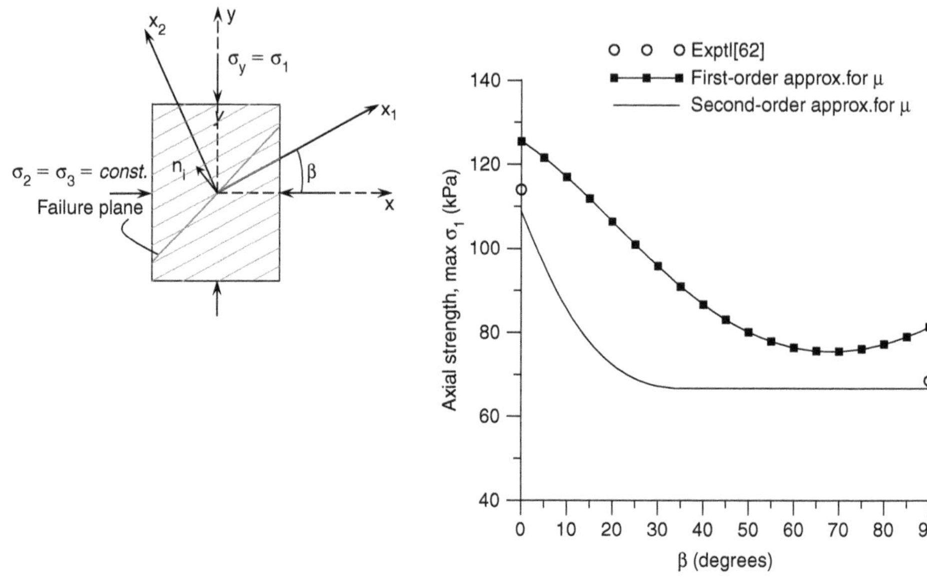

Figure 7.6 Variation of compressive strength with deposition angle β; triaxial tests at confinement of $p_0 = 10\,\text{kPa}$ (crushed limestone sand)

7.1.2 Formulation of failure criteria incorporating a microstructure tensor

The approach outlined here employs a scalar anisotropy parameter(s) which is defined in terms of mixed invariants of stress and microstructure tensors (Ref. [52]). The microstructure tensor a_{ij} is a measure of material fabric. While different descriptors may be employed to quantify the fabric (arrangement of intergranular contacts, pore size distribution, distribution of cracks, etc.), the eigenvectors of this operator, $e_i^{(\alpha)}$, $\alpha = 1, 2, 3$, define the preferred material axes. The spectral decomposition of a_{ij} takes the form

$$a_{ij} = a_1 e_i^{(1)} e_j^{(1)} + a_2 e_i^{(2)} e_j^{(2)} + a_3 e_i^{(3)} e_j^{(3)} = a_1 m_{ij}^{(1)} + a_2 m_{ij}^{(2)} + a_3 m_{ij}^{(3)} \tag{7.28}$$

where $m_{ij}^{(\alpha)} = e_i^{(\alpha)} e_j^{(\alpha)}$ are the respective structure-orientation tensors.

The most general approach assumes that the failure function F is an isotropic function of both σ_{ij} and a_{ij}, so that

$$F = F(\sigma_{ij}, a_{ij}) = F(T_{ip} T_{jq} \sigma_{pq}, T_{ip} T_{jq} a_{pq}) \tag{7.29}$$

where T_{ij} is the transformation tensor. Given the representation (7.29), the failure criterion can be expressed in terms of basic invariants of both operators as well as the respective joint invariants [57,58].

Clearly, the general form (7.29) is complex for practical engineering applications, particularly in the context of specification of material functions/parameters involved. Here, a simplified approach is followed that incorporates anisotropy measures which depend on relative orientation of principal axes of stress and microstructure tensor. Those descriptors are later identified with strength parameters (such as cohesion, angle of friction, etc.), so that, by analogy to the framework described in the preceding section, the strength properties are assumed to be orientation-dependent.

In order to define the anisotropy parameter(s), the formulation employs a generalized *loading vector* that is defined as

$$L_i = L_j e_i^{(j)}; \quad L_j = (\sigma_{j1}^2 + \sigma_{j2}^2 + \sigma_{j3}^2)^{1/2}; \quad (i,j = 1,2,3) \tag{7.30}$$

Thus, the components of L_i represent the magnitudes of traction vectors on the planes normal to the principal material axes. Note that

$$L_i^2 = e_k^{(i)} \sigma_{kj} e_l^{(i)} \sigma_{lj} = \mathrm{tr}\,(m_{kp}^{(i)} \sigma_{ql} \sigma_{lk}); \qquad L_k L_k = \sigma_{kj} \sigma_{kj} = \mathrm{tr}(\sigma_{kl} \sigma_{lj}) \tag{7.31}$$

so that the traction moduli can be expressed as mixed invariants of the stress and microstructure-orientation tensors.

The unit vector along L_i is now given by

$$l_i = \frac{L_i}{(L_k L_k)^{1/2}} = \left[\frac{(e_k^{(i)} \sigma_{kj} e_m^{(i)} \sigma_{mj})}{\sigma_{pq} \sigma_{pq}} \right]^{\frac{1}{2}} \tag{7.32}$$

and the projection of the microstructure tensor on l_i, becomes

$$\eta = a_{ij} l_i l_j = \frac{a_{ik} \sigma_{ij} \sigma_{kj}}{\sigma_{pq} \sigma_{pq}} \tag{7.33}$$

The scalar variable η, referred to as anisotropy parameter, specifies the effect of load orientation relative to material axes and is defined as the ratio of joint invariant of stress and microstructure tensor $a_{ij}\sigma_{ik}\sigma_{jk}$ to the stress invariant $\sigma_{ij}\sigma_{ij}$. It is a homogeneous function of stress of the degree zero, so that the stress magnitude does not affect its value. Note that eq.(7.33) can be expressed as

$$\eta = \eta_0 (1 + A_{ij} l_i l_j) \tag{7.34}$$

where

$$A_{ij} = \frac{1}{\eta_0} a_{ij} - \delta_{ij}; \quad \eta_0 = \frac{1}{3} a_{kk} \tag{7.35}$$

and $A_{ij} = \mathrm{dev}(a_{ij})/\eta_0$ is a symmetric traceless operator. The relation (7.34) can be generalized by considering higher order tensors, i.e.

$$\eta = \eta_0 (1 + A_{ij} l_i l_j + A_{ijkl} l_i l_j l_k l_l + \cdots) \tag{7.36}$$

Note that the functional form (7.36) is similar to that of representation (7.3). Thus, incorporating the same assumptions as those introduced in eq.(7.4), a simplified form can be employed, viz.

$$\eta = \eta_0(1 + A_{ij}l_i l_j + b_1(A_{ij}l_i l_j)^2 + b_2(A_{ij}l_i l_j)^3 + \cdots) \tag{7.37}$$

where b's are constants.

In view of the considerations above, the failure function (7.29) can be expressed in the general form

$$F = F(\sigma_{ij}, a_{ij}) = F(I_1, I_2, I_3, \eta) \tag{7.38}$$

where I_1, I_2 and I_3 are the basic invariants of the stress tensor. As mentioned earlier, the parameter η is typically identified with a relevant strength descriptor, whose value is then assumed to depend on the orientation of the sample relative to the direction of loading. Thus, the existing criteria can be easily extended to anisotropic regime by assuming that the strength parameters vary according to (7.37). In general, several different descriptors, and thus anisotropy parameters, may be employed simultaneously, depending on the type of material and the specific criterion used. Some illustrative examples are provided below.

(i) Mohr-Coulomb criterion

The Mohr-Coulomb criterion (see Sect. 2.4.1 and 3.1.1) can be expressed in the functional form

$$F = \sqrt{3}\bar{\sigma} - \eta_f g(\theta)(\sigma_m + C) = 0 \tag{7.39}$$

where

$$g(\theta) = \frac{3 - \sin\phi}{2\sqrt{3}\cos\theta - 2\sin\theta\sin\phi}; \quad \eta_f = \frac{6\sin\phi}{3 - \sin\phi}; \quad C = c\cot\phi \tag{7.40}$$

and ϕ, c are the angle of friction and cohesion, respectively.

An extension to the case of inherent anisotropy can be accomplished by assuming that the strength descriptors, in this case η_f and C, are orientation-dependent and have the representation analogous to that of (7.37). Note that, in the context of the criterion (7.39), C is the strength under hydrostatic tension. The latter, for an oriented structure, is in fact invariant with respect to orientation of the sample. Consider, for instance, the micro-structural arrangement used in examples given in the preceding section (viz. Figures 7.1–7.6). Referring the problem to the principal material axes x_i (so that $A_{ij} = 0$ for $i \neq j$, $A_{ii} = 0$), and noting that for hydrostatic tension, $\sigma_{ij} = \delta_{ij}p_0$, the components of the loading vector (7.32) become independent of the orientation β, i.e. $l_1 = l_2 = l_3 = 1/\sqrt{3} = const.$, there is $A_{ij}l_i l_j = (A_1 + A_2 + A_3)/3 \equiv 0$, which implies $C = const.$ Thus, the effects of anisotropy can be primarily attributed to variation in the strength parameter η_f, i.e.

$$\eta_f = \hat{\eta}_f(1 + A_{ij}l_i l_j + b_1(A_{ij}l_i l_j)^2 + b_2(A_{ij}l_i l_j)^3 + \cdots); \qquad C = const. \tag{7.41}$$

The specification of coefficients appearing in (7.41) entails the information on conditions at failure in samples tested at arbitrary orientation β with respect to direction of loading. Consider, for this purpose, the response of the sample under axial compression at a confining pressure p_0. Refer the geometry of the sample to the coordinate system identical to that employed before (e.g., Figure 7.6), whereby $x_i = \{x_1, x_2, x_3\}$ are the material axes. In this case,

$$A_{ij} l_i l_j = A_1(1 - 3l_2^2); \qquad l_2^2 = \frac{p_0^2 \sin^2\alpha + \sigma_1^2 \cos^2\alpha}{2p_0^2 + \sigma_1^2} \tag{7.42}$$

so that the function $\eta_f = \eta_f(l_i)$, eq.(7.41), reduces to

$$\eta_f = \hat{\eta}_f(1 + A_1(1 - 3l_2^2) + a_1 A_1^2(1 - 3l_2^2)^2 + a_2 A_1^3(1 - 3l_2^2)^3 \\ + a_3 A_1^4(1 - 3l_2^2)^4 + ...) \tag{7.43}$$

Note that in the absence of confinement, $p_0 = 0$, there is $l_2 = \cos\beta$, so that the anisotropy parameter is an explicit function of the deposition angle β, i.e.

$$l_2^2 = \cos^2\alpha \;\Rightarrow\; \eta_f = \hat{\eta}_f(1 + A_1(1 - 3\cos^2\beta) + b_1 A_1^2(1 - 3\cos^2\beta)^2 + ...) \tag{7.44}$$

As mentioned earlier, 'triaxial' tests on inclined samples, which are required for the identification purposes, cannot be easily conducted due to the limitations of the 'triaxial' equipment. Thus, an implicit approach is adopted here, whereby the key information is acquired using the predictions based on the critical plane approach. As an illustration, consider the response of crushed limestone sand, for which the mechanical properties are the same as those used in the last example of the preceding section (Figures 7.5–7.6).

The identification procedure employs the predictions of compressive strength at confinements of $p_0 = 0$ and 10 kPa. The latter are based, for simplicity, on the first-order approximations to μ and c, as depicted in Figure 7.5. Note that the results for the confinement of 10 kPa are the same as those shown in Figure 7.6. The value of the strength parameter C, which is required to evaluate η_f for the given stress state, was estimated by examining the response under hydrostatic tension σ_t. In this case, for any orientation n_i there is $\tau = 0$, $\sigma = \sigma_t$. Thus, the constrained optimization problem (7.5), employing representation (7.19), becomes

$$\max F = \underbrace{\max}_{n_i}(\sigma_t - c/\mu) = 0 \quad\Rightarrow\quad \sigma_t = C = \underbrace{\min}_{n_i}(c/\mu) = const \tag{7.45}$$

Here, using the data presented in Figure 7.5, there is $C = \min(c/\mu) = 2.4$ kPa (at $\beta = 0$).

Given now the stress parameters $\{p, q\}$ at failure, for each specific orientation of the sample, the anisotropy parameter can be determined, i.e. $\eta_f = q/(p + C)$, together with the corresponding value of l_2, eq.(7.42). The results can then be plotted in the affined space $\{\eta_f, l_2\}$ to obtain a set of data describing the relation (7.43).

To demonstrate the procedure, Figure 7.7a shows the critical plane predictions for the conditions at failure in samples tested at $\beta = 0°$ and $\beta = 90°$, at both confining pressures $p_0 = 0$, 10 kPa. At the same time, Figure 7.7b gives the corresponding relation

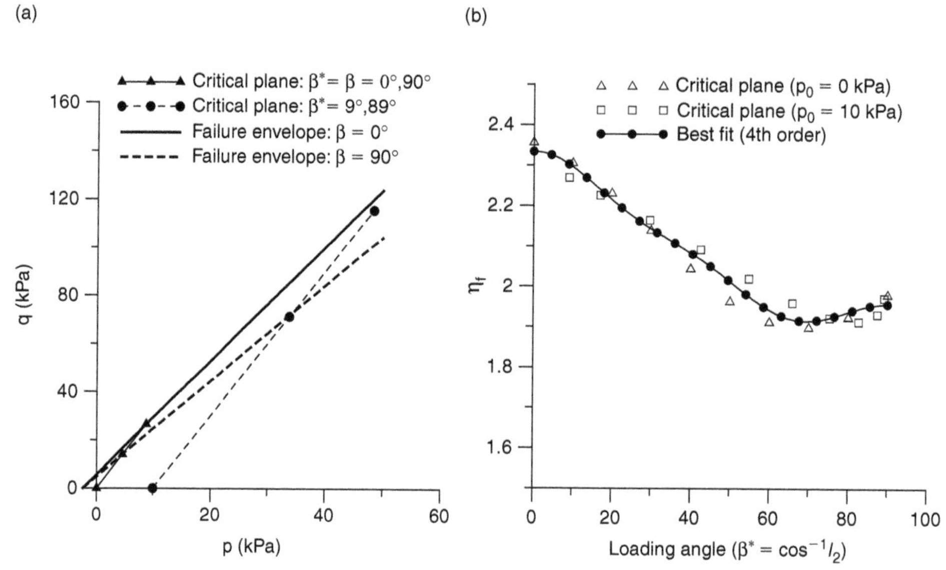

Figure 7.7 (a) Failure envelopes for $\beta = 0°, 90°$ based on critical plane predictions; (b) Variation of the strength parameter η_f with the loading angle (crushed limestone sand)

between η_f and the loading angle $\beta^* = \cos^{-1} l_2$. It should be noted that, in general, the results for $p_0 = 0$, which correspond to $\beta^* = \beta$ (i.e. η_f is a direct function of the deposition angle), are sufficient to identify the coefficients $\hat{\eta}_f$, A_1, b_1, $b_2, \ldots$, viz. eq.(7.44). A better representation though is obtained if these are augmented by the results corresponding to other confining pressures. Here, the best-fit incorporating the terms up to the order four in representation (7.43) is obtained using both sets of data, i.e. for $p_0 = 0$ and $p_0 = 10$ kPa. The resulting values of the coefficients of approximation function are

$$\hat{\eta}_f = 1.9758,\ A_1 = -0.07563,\ b_1 = 2.2450,\ b_2 = -83.978,\ b_3 = 514.04$$

Once the variation of the anisotropy parameter is defined, the framework can be used in a predictive mode. As an example, Figure 7.8 shows the evolution of strength at confinements of $p_0 = 0$, 5 kPa and 10 kPa, as obtained from representation (7.39). The results are compared with predictions based on critical plane approach. It is evident that both qualitative and quantitative aspects of these predictions are very similar.

(ii) Tresca/von Mises criterion

The Tresca criterion is a particular case of representation (7.39), which corresponds to $\phi = 0$. Thus,

$$F = \sqrt{3}\bar{\sigma} - g(\theta)\sigma_Y = 0; \quad g(\theta) = \frac{3}{2\sqrt{3}\cos\theta} \tag{7.46}$$

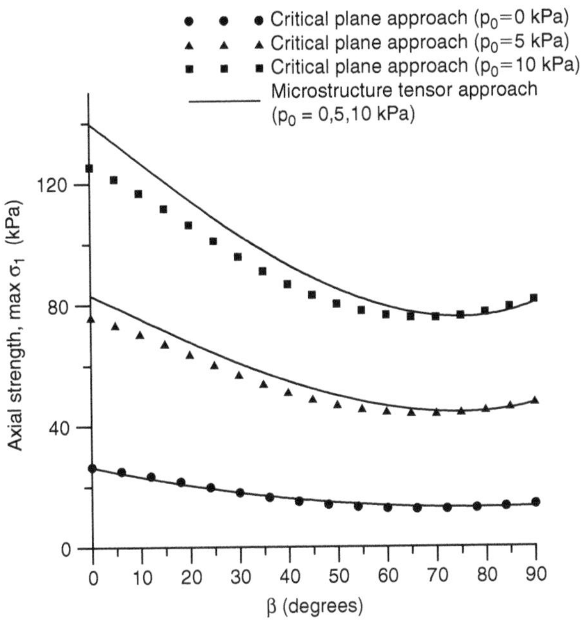

Figure 7.8 Evolution of compressive strength with deposition angle β; comparison with critical plane approach (crushed limestone sand)

where $\sigma_Y = 2c$ is the ultimate stress in axial tension/compression, often identified with the yield stress. Note that for $g(\theta) = const.$, the representation (7.46) reduces to von Mises criterion (see Sect. 2.4.2).

Again, an extension to the case of inherent anisotropy involves an assumption that the strength descriptor σ_Y is orientation-dependent, so that

$$\sigma_Y = \hat{\sigma}_Y(1 + A_{ij}l_il_j + b_1(A_{ij}l_il_j)^2 + b_2(A_{ij}l_il_j)^3 + \cdots) \tag{7.47}$$

The identification of anisotropy function (7.47) requires tests on differently oriented samples subjected to, for example, uniaxial tension. Referring the problem to the same frame of reference as before (cf. Figure 7.6), σ_Y can be expressed in the form analogous to (7.44), i.e.

$$\sigma_Y = \hat{\sigma}_Y(1 + A_1(1 - 3\cos^2\beta) + b_1A_1^2(1 - 3\cos^2\beta)^2 + \cdots) \tag{7.48}$$

In order to illustrate the identification procedure, consider again the predictions based on the critical plane approach as the source of data. In this case, the representation (7.7) together with (7.10) can be employed, which corresponds to a linear approximation

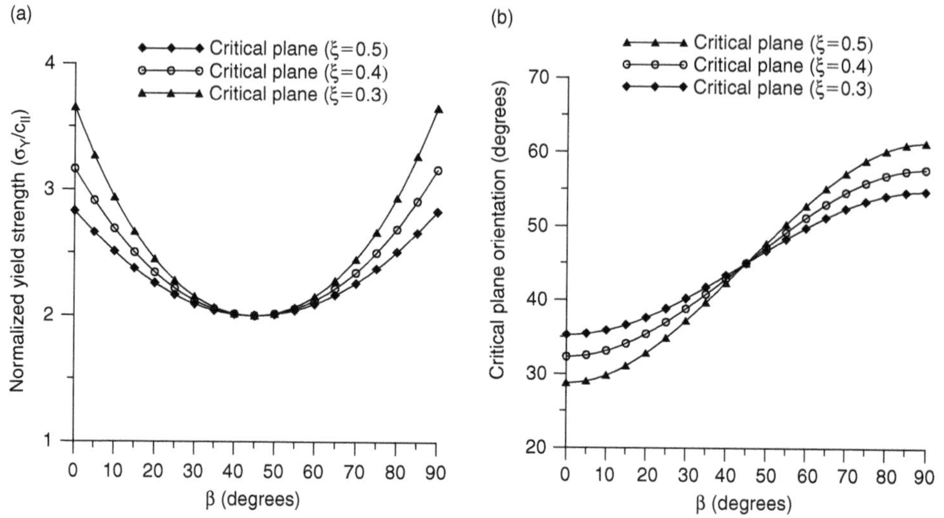

Figure 7.9 Variation of normalized tensile strength and orientation of the critical plane with preferred orientation β; axial tension (Tresca material)

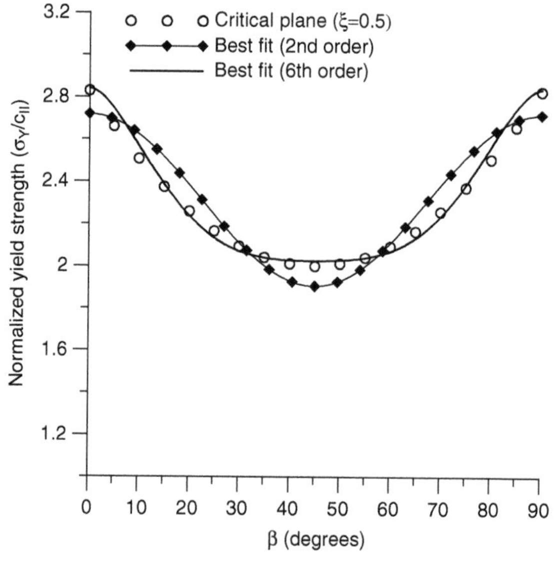

Figure 7.10 Identification of anisotropy function $\sigma_Y = \sigma_Y(\beta)$, eq.(7.46), based on predictions incorporating critical plane approach (Tresca material)

for the shear strength $c(n_i)$. Assume that the value of c at $\beta = 0°$ and $\beta = 90°$ is $c_\parallel$ and $c_\perp$, respectively and that $c_\parallel / c_\perp = \xi$. Thus, in accordance with eq.(7.24) or (7.26),

$$c_\parallel / c_\perp = \xi; \quad c_0 = c_\parallel \left(\frac{2+\xi}{3\xi} \right); \quad \Omega_1 = \frac{1-\xi}{2+\xi}$$

Figure 7.9a shows the predicted variation of normalized tensile strength with preferred orientation β, for different values of ξ. At the same time, Figure 7.9b gives the orientation of the critical/localization plane as a function of β.

The above results can now be employed to identify the anisotropy function (7.48). Figure 7.10 shows the best-fit approximations to the distribution corresponding to $\xi = 0.5$. The 2nd and the 6th order fits are provided that identify the coefficients embedded in representation (7.47). Evidently, the approximation including the terms up to the order six is more accurate and corresponds to

$$\hat{\sigma}_Y = 2.0516,\ A_1 = 0.079798,\ b_1 = 1.8602,\ b_2 = -146.64,\ b_3 = 2866.3,$$
$$b_4 = 56920,\ b_5 = 237770$$

Again, given the set of coefficients above the criterion (7.46) can be employed in a predictive mode.

7.2 DESCRIPTION OF INELASTIC DEFORMATION PROCESS

The focus so far has been on specification of the conditions at failure in the presence of inherent anisotropy. In this part, the methodologies outlined in Sect. 7.1.1 and 7.1.2 are extended to incorporate the description of the deformation process. Both, the critical plane approach as well as the framework incorporating a microstructure tensor are employed and the corresponding plasticity formulations are discussed.

7.2.1 Plasticity formulation for critical plane approach

Within this framework, the inelastic deformation is attributed to sliding/separation along an infinite set of randomly oriented planes. For each plane, the conditions at failure are represented by the local criterion $F = 0$, where F is defined viz. eq.(7.1) and incorporates the scalar-valued function (7.3) or (7.4). The irreversible deformation is accounted for by invoking an appropriate plasticity framework. This approach is conceptually similar to the so-called multi-laminate framework (cf. Refs. [63,64]).

The yield/loading surface, as well as the plastic potential, both are defined in terms of the normal and tangential components of the traction vector acting on a plane with a unit normal n_i, i.e.

$$f(n_i) = f(\sigma, \tau, \kappa) = 0; \qquad \psi(n_i) = \psi(\sigma, \tau) = const \tag{7.49}$$

where κ is a hardening parameter, which is a function of the plastic deformation history. The equation of the loading surface is formulated in such a manner that $\kappa \to \infty \Rightarrow f \to F$, where F is the respective failure function.

Referring the problem to a local frame of reference $\bar{x}_i$ associated with the base vectors $t_i^s/|t_i^s|$ and n_i, the flow rule may be written as

$$d\bar{e}_i^p = d\lambda \frac{\partial \psi}{\partial \bar{t}_i} \tag{7.50}$$

where $\bar{e}_i$ is the strain vector, whereas the corresponding traction $\bar{t}_i$ has the components $\bar{t}_i = \{\tau, \sigma, 0\}$. The strain rates contributed by this plane are expressed as a symmetric part of a dyadic product

$$d\varepsilon_{ij}^p = \frac{1}{2}(de_i^p\, n_j + de_j^p n_i); \qquad de_i^p = T_{ij}\, d\bar{e}_j^p \tag{7.51}$$

Here, x_i is an arbitrary global frame of reference and T_{ij} is the transformation matrix. Thus, substituting (7.50) into (7.51)

$$d\varepsilon_{ij}^p = \frac{1}{2} d\lambda (T_{ik} n_j + T_{jk} n_i) \frac{\partial \psi}{\partial \bar{t}_k} \tag{7.52}$$

The overall macroscopic deformation is now obtained by averaging the contributions from all active planes, so that

$$d\varepsilon_{ij}^p = \frac{1}{8\pi} \int_S \left\{ d\lambda (T_{ik} n_j + T_{jk}\, n_i) \frac{\partial \psi}{\partial \bar{t}_k} \right\} dS \tag{7.53}$$

where the integration is carried out over a surface area (S) of the unit sphere. In practical implementations, the integration process is performed numerically by adopting a set of 'sampling planes'. Details concerning the orientation of these planes and the distribution of weight coefficients are provided in Ref. [63]. Given the representation (7.53), the macroscopic constitutive relation can be obtained by invoking the additivity postulate, i.e.

$$d\varepsilon_{ij} = C_{ijkl}^e d\sigma_{kl} + d\varepsilon_{ij}^p = C_{ijkl}^e d\sigma_{kl} + \frac{1}{8\pi} \int_S \left\{ d\lambda (T_{ik}\, n_j + T_{jk} n_i) \frac{\partial \psi}{\partial \bar{t}_k} \right\} dS \tag{7.54}$$

where C_{ijkl}^e is the elastic compliance operator, whose functional form is affected by the presence of anisotropy.

As an illustration, consider the framework of deviatoric hardening, similar to that outlined in Chapter 3. In particular, assume that the loading surface is of the form

$$f(n_i) = \tau + \eta\sigma - c = 0; \quad \eta = \eta(\kappa) = \mu \frac{\kappa}{A + \kappa}; \quad \kappa = \int |d\gamma^p| \tag{7.55}$$

Here, $\mu = \mu(n_i)$, $c = c(n_i)$ assume the value stipulated by representation (7.4), while $d\gamma^p = d\bar{e}_1^p$ and A is a material constant. It should be noted again that for $\kappa \to \infty$ there is $\eta \to \mu$ which implies that $f(n_i) \to F(n_i)$, where F is the failure function consistent with Coulomb's criterion.

The plastic deformation can be described by invoking a non-associated flow rule, eq.(7.50), in which the plastic potential is defined as

$$\psi(n_i) = \tau - \mu_c(\sigma - \sigma_0) \ln \frac{\sigma_0 - \sigma}{\sigma^0} = 0; \quad \sigma_0 = c/\mu \tag{7.56}$$

where σ^0 is evaluated from the condition of $\psi(n_i)=0$, whereas $\mu_c \propto \mu$ is a parameter which represents the value of $\eta=\tau/(\sigma_0-\sigma)$ at which a transition from compaction to dilatancy takes place.

It should be noted that the elastic properties associated with each sampling plane can be defined by invoking the dyadic decomposition in eq.(7.51). Thus,

$$t_i = \sigma_{ij} n_j = B_{ik} e_k; \quad B_{ik} = D_{ijkl} n_j n_l \tag{7.57}$$

where $D_{ijkl} = C^{-1}_{ijkl}$ is the elastic stiffness operator. The constitutive relation can now be obtained following the standard plasticity procedure as outlined in Chapter 3.

7.2.2 Plasticity formulation incorporating a microstructure tensor

The general plasticity formulation can be derived by assuming the yield/loading surface in the form consistent with representation (7.38), i.e.

$$f = f(\sigma_{ij}, a_{ij}, \varepsilon^p_{ij}) = f(I_1, I_2, I_3, \eta, \kappa) = 0 \tag{7.58}$$

where κ is a scalar-valued function of plastic deformation and η is the anisotropy parameter, eq.(7.33). The flow rule may now be written as

$$d\varepsilon^p_{ij} = d\lambda \frac{\partial \psi}{\partial \sigma_{ij}}; \quad \psi = \psi(\sigma_{ij}, a_{ij}) = \psi(I_1, I_2, I_3, \eta) = const. \tag{7.59}$$

where ψ is the plastic potential function.

A specific formulation can be obtained by a suitable generalization of the classical isotropic criteria. As an illustration, consider again the framework of deviatoric hardening. Within this framework, the loading surface can be defined by extending the representation (7.39) to the form

$$f = \sqrt{3}\bar{\sigma} - \vartheta g(\theta)(\sigma_m + C) = 0; \quad \vartheta = \vartheta(\kappa) = \eta_f \frac{\kappa}{B+\kappa}; \quad d\kappa = d\varepsilon^p_q = \sqrt{J_{2\dot{\varepsilon}}} \tag{7.60}$$

where $J_{2\dot{\varepsilon}}$ is the second invariant of the deviatoric part of the plastic strain increment (cf. Sect. 3.3.3.) and B is a material constant. Note, once again, that for for $\kappa \to \infty$ there is $\vartheta \to \eta_f$ which implies that $f \to F$, so that the conditions at failure are consistent with Mohr-Coulomb criterion (7.39). It is also noted that, according to representation (7.41), $\eta_f = \eta_f(l_i)$ is orientation-dependent, while $C = const.$

The plastic potential can be chosen as

$$\psi = \sqrt{3}\bar{\sigma} + \eta_c g(\theta)(\sigma_m + C) \ln \frac{(\sigma_m + C)}{\sigma^0_m} = 0 \tag{7.61}$$

where $\mu_c \propto \eta$, so that $\eta_c = \eta_c(l_i)$. Following now the standard plasticity procedure (cf. Sect. 3.3.3), i.e. invoking the consistency condition $df = 0$, yields

$$d\varepsilon_{ij}^p = d\lambda \frac{\partial\psi}{\partial\sigma_{ij}}; \quad d\lambda = H_p^{-1}\frac{\partial f}{\partial\sigma_{ij}} d\sigma_{ij}; \; H_p = -\sqrt{\frac{2}{3}}\frac{\partial f}{\partial\vartheta}\frac{\partial\vartheta}{\partial\varepsilon^p}\left(dev\frac{\partial\psi}{\partial\sigma_{ij}}\, dev\frac{\partial\psi}{\partial\sigma_{ij}}\right)^{\frac{1}{2}} \tag{7.62}$$

where H_p is the plastic hardening modulus.

For the functional form (7.60), the gradient operator can be expressed as

$$\frac{\partial f}{\partial\sigma_{ij}} = \left(\frac{\partial f}{\partial\sigma_m}\frac{\partial\sigma_m}{\partial\sigma_{ij}} + \frac{\partial f}{\partial\bar{\sigma}}\frac{\partial\bar{\sigma}}{\partial\sigma_{ij}} + \frac{\partial f}{\partial\theta}\frac{\partial\theta}{\partial\sigma_{ij}}\right) + \left(\frac{\partial f}{\partial\eta_f}\frac{\partial\eta_f}{\partial\sigma_{ij}}\right) \tag{7.63}$$

where (cf. Sect. 2.5)

$$\frac{\partial\sigma_m}{\partial\sigma_{ij}} = -\frac{1}{3}\delta_{ij}; \; \frac{\partial\bar{\sigma}}{\partial\sigma_{ij}} = \frac{1}{2\bar{\sigma}}s_{ij}; \; \frac{\partial\theta}{\partial\sigma_{ij}} = \frac{\sqrt{3}}{2\bar{\sigma}^3\cos 3\theta}\left(\frac{3J_3}{2\bar{\sigma}^2}s_{ij} - s_{ik}s_{kj} + \frac{2}{3}\bar{\sigma}^2\delta_{ij}\right) \tag{7.64}$$

In this case, the effect of anisotropy is embedded in the last term of eq.(7.63). In order to evaluate this expression, note that according to eq.(7.60)

$$\frac{\partial f}{\partial\eta_f} = -\vartheta(\kappa)\,g(\theta)\sigma_m \tag{7.65}$$

whereas

$$\frac{\partial\eta_f}{\partial\sigma_{ij}} = \frac{\partial\eta_f}{\partial\zeta}\frac{\partial\zeta}{\partial\sigma_{ij}}; \qquad \zeta = A_{ij}l_il_j = \frac{A_{ik}\sigma_{ij}\sigma_{kj}}{\sigma_{pq}\sigma_{pq}} \tag{7.66}$$

Substituting for η_f from eq.(7.41) and differentiating, one obtains (in view of symmetry of A_{ij})

$$\frac{\partial\eta_f}{\partial\sigma_{ij}} = 2\hat{\eta}_f(1 + 2b_1\zeta + 3b_2\zeta^2 + \cdots)\frac{A_{ki}\sigma_{kj}\sigma_{pq}\sigma_{pq} - A_{pk}\sigma_{pq}\sigma_{kq}\sigma_{ij}}{(\sigma_{mn}\sigma_{mn})^2} \tag{7.67}$$

which completes the specification of the gradient tensor in eq.(7.63). Note that a similar representation can be obtained for the gradient of the plastic potential function.

Given the functional form of both gradient operators, the constitutive relation can now be obtained by invoking the additivity postulate, i.e.

$$d\varepsilon_{ij} = C_{ijkl}^e d\sigma_{kl} + d\varepsilon_{ij}^p = \left(C_{ijkl}^e + H_p^{-1}\frac{\partial f}{\partial\sigma_{ij}}\frac{\partial\psi}{\partial\sigma_{kl}}\right)d\sigma_{kl} = C_{ijkl}d\sigma_{kl} \tag{7.68}$$

where C_{ijkl}^e is the elastic compliance whose representation, once again, depends on the type of material anisotropy.

7.2.3 Numerical examples

As an example, consider again the response of crushed limestone sand subjected to axial compression at different deposition angles β. The basic material parameters defining the conditions at failure for both the critical plane and the microstructure tensor formulations , i.e. (7.19) and (7.39) respectively, are the same as those given in examples provided in the preceding section 7.1.2(*i*). Thus, for the critical plane formulation, the first-order approximations for μ and c are employed, viz. eqs.(7.24) and (7.26), which correspond to

$$\mu_0 = 1.3,\ \Omega_1^{\mu} = 0.1; \quad c_0 = 3.76\,\text{kPa},\ \Omega_1^{c} = 0.17$$

For the framework incorporating the microstructure tensor, the terms up to the order of 4 are retained in (7.43) and the values of the coefficients are identical to those given in Sect. 7.1.2(*i*), i.e.

$$\hat{\eta}_f = 1.9758,\ A_1 = -0.07563,\ b_1 = 2.2450,\ b_2 = -83.978,\ b_3 = 514.04$$

The specific problem addressed here is the evolution of deformation characteristics in samples tested at the confinement of $p_0 = 5$ kPa. The strength characteristics for this case, which correspond to the set of parameters listed above, are provided in Figure 7.8. The problem is referred again to the same global frame of reference $\{x,y,z\}$ as before, while x_2 is the preferred material axis (cf. Figures 7.1–7.6).

The plasticity formulations outlined in Sect. 7.2.1 and 7.2.2 require the information on the material parameters A, μ_c (viz. eqs.(7.55) and (7.56)) as well as B, η_c (viz. eqs.(7.60) and (7.61)). The simulations presented here have been carried out assuming $\{A=0.01,\ \mu_c=0.9\mu\}$ for the critical plane approach and $\{B=0.001,\ \eta_c=0.9\eta_f\}$ for the microstructure tensor formulation. The values were assigned rather arbitrarily as no experimental information is available. The primary objective was to compare the basic trends in mechanical response as predicted by both these approaches. The values of elastic constants were also chosen on intuitive basis, as

$$E_1 = 10\,\text{MPa},\ E_2 = 20\,\text{MPa},\ \nu_{13} = 0.25,\ \nu_{21} = 0.20,\ G_{12} = 8\,\text{MPa}$$

The simulations were carried out for $\beta=0°$, 45°, and 90° using the constitutive relations (7.54) and (7.68). For the multi-laminate framework, the integration was carried out using a set of 33 sampling planes distributed in 3D-space. Details concerning the orientation of these planes and the respective values of weigh coefficients are provided in Ref.[63] cited earlier.

The main results of numerical simulations are presented in Figure 7.11. Figure 7.11a shows the deviatoric stress-strain characteristics for different deposition angles β. The predictions from both the critical plane and the microstructure tensor approach are given. It should be mentioned that the deformation mode is axisymmetric only for the sample tested at $\beta=0°$. For $\beta=90°$, the lateral deformation is different in x and y directions, while for sample at $\beta=45°$, the shear strains develop. Thus, the deviatoric strain in Figure 7.11 is defined using the general form provided in eq.(7.60). Figure 7.11b shows the corresponding volume change characteristics. For both approaches, there is a progressive transition from compaction, at the early stages of the deformation process, to dilatancy and the quantitative results are sensitive to the orientation of the sample.

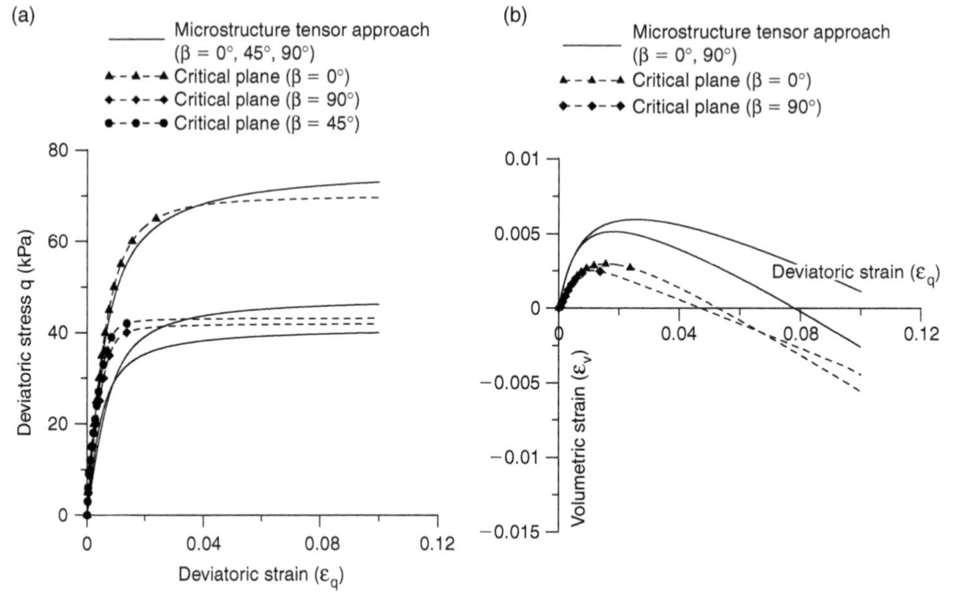

Figure 7.11 Comparison of material characteristics in triaxial compression ($p_0 = 5$ kPa) at different deposition angles β; (a) deviatoric stress-strain response, (b) evolution of volume change (crushed limestone sand)

It is evident from the example given above that both approaches, viz. (7.54) and (7.68), predict similar trends in the mechanical response of the material. The results capture the sensitivity of the strength, as well as the deformation mode, to the orientation of the sample. The predicted trends are consistent with the experimental evidence, as reported for example in Ref. [62]. The choice of the approach now largely depends on availability of experimental data for specification of material constants as well as the numerical efficiency. In the context of identification of parameters, the formulation based on the critical plane approach requires a series of direct shear tests on samples at different orientation relative to the preferred directions of microstructure. For soils such tests can be readily performed, they are not very common though for other geomaterials. The approach incorporating a microstructure tensor requires uniaxial/'triaxial' tests on differently oriented samples. While the uniaxial tests on cohesive materials present no difficulty, the results of 'triaxial' tests on inclined specimens are not reliable as the experimental setup imposes kinematic constraints that trigger a non-uniform deformation. Thus, modified setups need yet to be developed to test arbitrarily oriented samples under multi-axial load. In terms of numerical implementation, the microstructure tensor approach appears to be at an advantage as it involves a relatively simple enhancement of the standard isotropic framework. The critical plane approach, on the other hand, requires implementation of an averaging scheme whereby the overall macroscopic deformation is obtained by examining the contributions from all active planes. Such an approach is computationally less efficient.

Chapter 8

General trends in the mechanical behaviour of soils and rocks

The primary objective of this final chapter is to provide the reader with an overview of the experimental response of geological media. This material supplements the considerations in the other parts of the book, in particular Chapters 3, 4 and 7, and provides an insight into the complexities associated with mechanical characteristics of geomaterials. The focus is on identifying the influence of key factors, such as the arrangement of internal structure, the loading configuration, etc.

It is evident from the discussion in the preceding chapter that the mechanical response of both soils and rocks is strongly influenced by the microstructure. Most soils have a random structure, implying that at the macro-level their behaviour is isotropic. In this case, scalar descriptors are typically employed, such as initial porosity/void ratio. On the other hand, as pointed out in Chapter 7, in a number of geomaterials the microstructure exhibits a strong inherent anisotropy. Examples here include granular soils with particles aligned with the plane of deposition, as well as sedimentary rocks; the latter formed by deposits of clay and silt sediments. The anisotropy is linked to the existence of preferred orientation, e.g. bedding planes that mark the limits of strata and can often be identified by a visual examination. For this class of materials, both the deformation and strength characteristics are orientation-dependent and the description of fabric requires some tensorial, rather than scalar descriptors (cf. Chapter 7).

Geomaterials are porous, so that their mechanical response is strongly affected by the confining pressure. This is in contrast to metals, where no such sensitivity is recorded, as pointed out in Chapter 2. In geomaterials, the tensile strength is, in general, marginal as compared to that in compression regime. Furthermore, all porous media display evolution of volume ranging from compaction to dilation that depends, once again, on the microstructural arrangement. Finally, in many practical geotechnical problems, the engineers need to deal with a two- or even three-phase material; i.e. soils/rocks are either fully or partially saturated with water. The presence of fluid alters the mechanical characteristics quite significantly. For granular materials, for example, such diversified effects as liquefaction and cyclic mobility can occur.

The aim of this chapter is to outline the basic trends in the mechanical behaviour of soils and rocks. Various loading scenarios are examined and both drained and undrained characteristics are reviewed. Later in this chapter, the identification procedure for material parameters employed in standard plasticity formulations, as discussed earlier in Chapter 3, is summarized.

8.1 BASIC MECHANICAL CHARACTERISTICS IN MONOTONIC TESTS UNDER DRAINED CONDITIONS

As mentioned earlier, the most common experimental tests for soils are the 'triaxial' compression and extension tests conducted under fully drained conditions. Typical loading configurations involve axial compression/extension and lateral compression/ extension, as discussed in Section 3.1.1. In all these tests, the samples are subjected to initial confining/cell pressure and failed by an increase/decrease in either the vertical or lateral compressive stress. Similar tests are also conducted on rocks though the most common ones involve the axial compression. In what follows, some typical results are shown, for both soils as well as rocks, to highlight the basic trends in their mechanical response. The focus in this section is on the influence of confining pressure and the evolution of volume change in monotonic tests that are carried out under drained conditions. The related issues, which include the influence of Lode's angle and the occurrence of localized deformation, are also addressed.

8.1.1 Influence of confining pressure; compaction/dilatancy

Drained 'triaxial' compression and extension tests are typically performed over a range of different confining pressures. For a porous microstructure, which is typical of all geomaterials, the confining pressure has a strong influence on the mechanical behaviour, affecting not only the conditions at failure but also the deformation characteristics.

Figures 8.1 and 8.2 show typical results of a series of drained 'triaxial' compression tests conducted on both loose and dense sand. The results are taken from the article by Kolymbas and Wu [65] and involve tests on Karlsruhe sand, which is a uniform material consisting of subround medium quartz grains with the maximum and minimum specific weights of 17 kN/m^3 and 14 kN/m^3, respectively. The tests were carried out at confining pressures ranging from $p_0 = 50$ kPa to $p_0 = 1000$ kPa.

The experimental results clearly indicate that the strength increases with the increase in confinement. In loose sand (relative density $D_r = 16\%$, Figure 8.1), the mechanical response is stable (in Drucker's sense), i.e. the failure is associated with unlimited deformation that develops as the ultimate load is approached (Figure 8.1a,b). An increase in the deviatoric stress is associated with progressive *compaction* (decrease in volume), Figure 8.1c. The compressibility of the sample increases with the increase in confining pressure, and as the conditions at failure are approached the volume change tends to zero. In dense specimens (cf. Figure 8.2, relative density $D_r = 98\%$), the strength is significantly higher and the stress-strain characteristic is typically associated with a strain-softening response (viz. descending branch in Figure 8.2a) at the macroscale. At the initial stage of the deformation process, the material undergoes compaction, Figure 8.2b. At higher deviatoric stress intensities though, a transition to *dilation* (i.e. increase in volume) occurs. The dilation is clearly more pronounced at lower confining pressures.

The results presented in Figures 8.1 and 8.2 allow to assess the conditions at failure in the 'triaxial' compression regime. Figure 8.3 shows the *p-q* representation of the failure envelopes corresponding to both loose and dense Karlsruhe sand. The stress states at which the failure commences are identified based on the deviatoric characteristics in Figures 8.1b and 8.2a, respectively. The results of experiments are

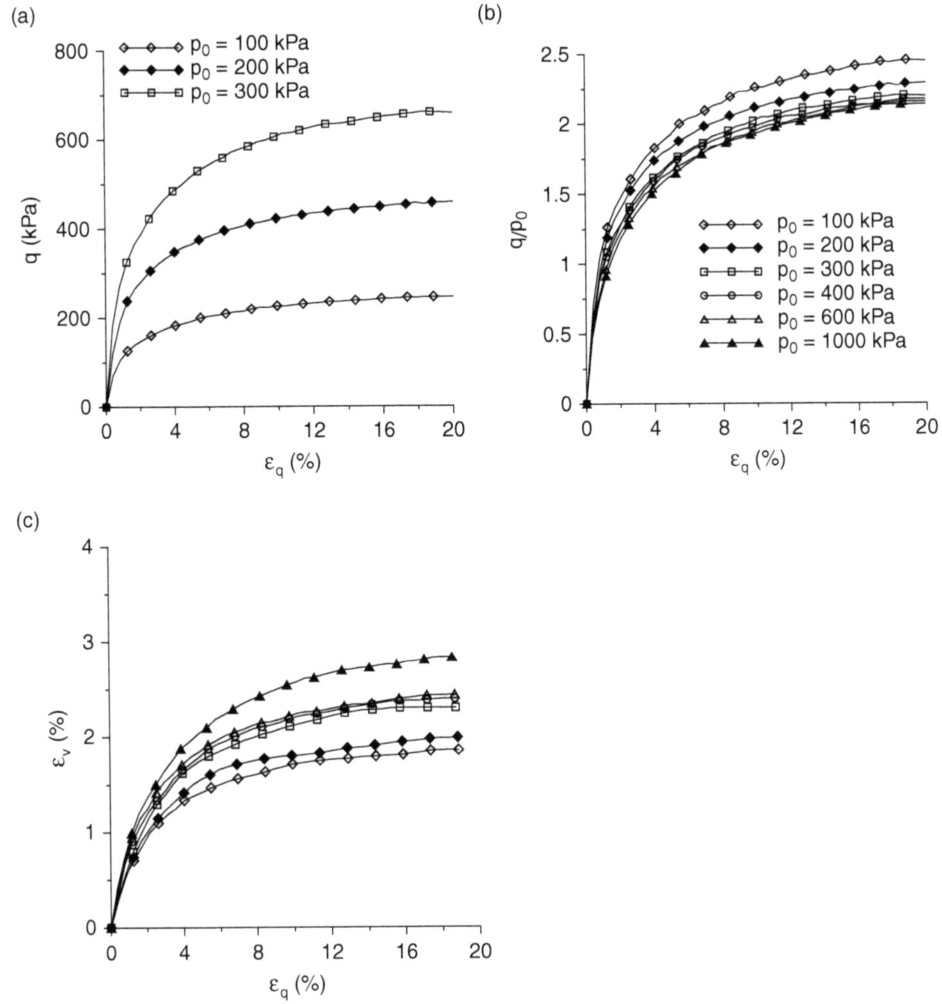

Figure 8.1 Results of triaxial tests on loose Karlsruhe sand (data from Ref. [65]); (a) deviatoric stress-strain characteristics, (b) normalized characteristics covering a broader range of confinement (c) evolution of volume change

approximated by a best fit line, which corresponds to Mohr-Coulomb representation. It is evident that this approximation is quite adequate here, implying that meridional sections of the failure envelope may be considered as linear over a broad range of confining pressures.

As a further illustration, another set of results for dense Sacramento river sand ($D_r \simeq 100\%$) is given in Figure 8.4. The data is taken from the article by Lade [66]; the testing configuration is similar to that examined in Figure 8.2, while the tests cover a broader range of strain for up to 25%. Figure 8.4a gives the deviatoric characteristics at confining pressures p_0 ranging from 100 kPa to nearly 2000 kPa, whereas Figure 8.4b

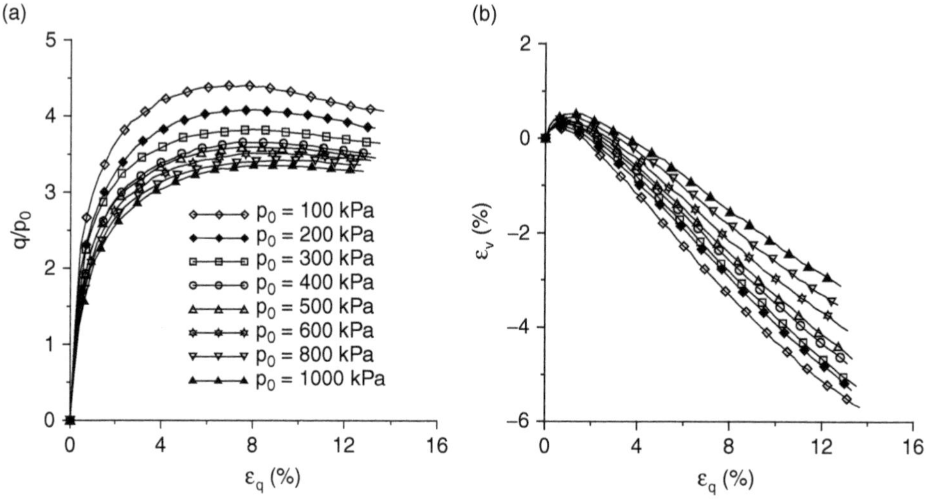

Figure 8.2 Results of triaxial tests on dense Karlsruhe sand (data from Ref. [65]); (a) normalized deviatoric characteristics, (b) evolution of volume change

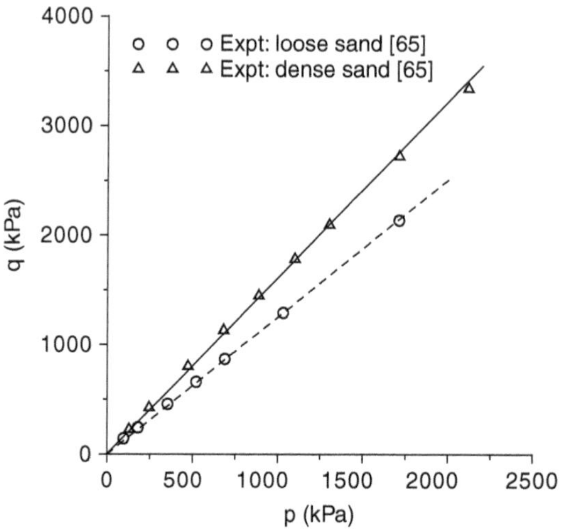

Figure 8.3 Mohr-Coulomb failure envelopes for loose and dense Karlsruhe sand

shows the corresponding evolution of volume change. The general trends are consistent with those depicted earlier in Figure 8.2; the strength increases with increasing confinement and there is clearly the evidence of unstable strain-softening response, which is typically associated with formation of a shear band. For the volume change characteristics, Figure 8.4b, it is interesting to note that the dilation, which stems from reorientation of grains, is quite pronounced at low confining pressures, while at high pressures it virtually disappears. Also, in the softening regime, the rate of dilation decreases and, as the residual state is approached, no volume change is recorded.

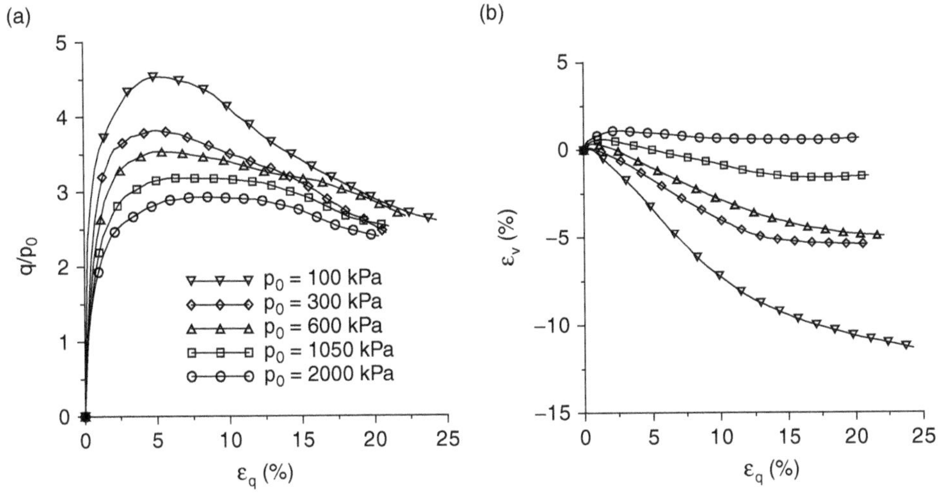

Figure 8.4 Results of triaxial tests on dense Sacramento River sand (data from Ref. [66]); (a) normalized deviatoric characteristics, (b) evolution of volume change

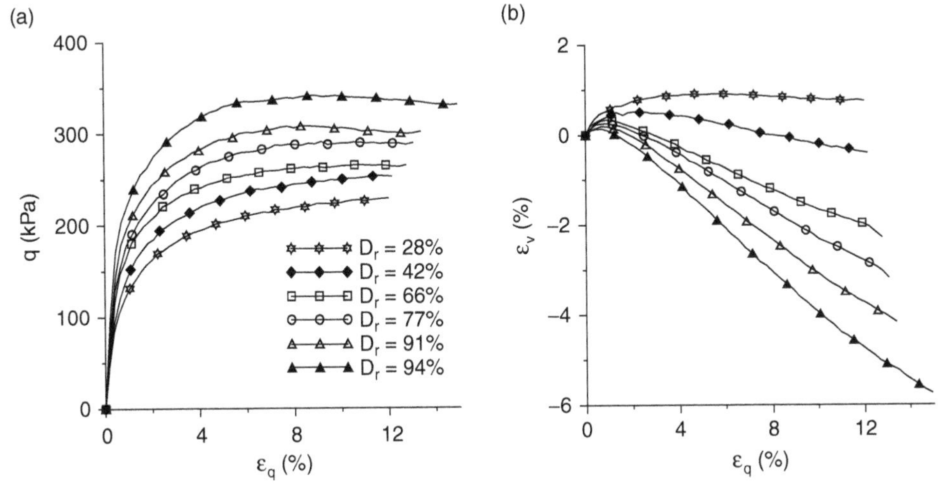

Figure 8.5 Triaxial tests on Karlsruhe sand at $p_0 = 100\,\mathrm{kPa}$ and different relative densities D_r (data from Ref. [65]); (a) deviatoric characteristics, (b) evolution of volume change

It is clear from the results presented above that the mechanical behaviour of sand is strongly affected by the confining pressure as well as the initial degree of compaction. The latter point is further elaborated in Figure 8.5, which provides the mechanical characteristics for Karlsruhe sand (Ref. [65]) at a fixed confinement of $p_0 = 100\,\mathrm{kPa}$ and different relative densities D_r, ranging from 28% to 94%. It is noted that the increase in initial density (i.e. transition from loose to dense state of compaction) results in an increase in the strength, and thus in the angle of internal friction ϕ. At the same time, the denser the material the more pronounced the dilation effects become, which is evidenced in Figure 8.5b.

Consider now the mechanical response of rocks under a similar loading environment. Let us restrict ourselves first to the type of rocks which have a homogeneous random microstructure, so that their behaviour at the macroscale is isotropic. In mechanical terms, the primary difference between granular media, such as sand, and rocks is the fact that the former pose no resistance to tension. Most rocks, on the other hand, are relatively hard, naturally formed aggregates of mineral matter and their tensile strength can be quite significant (with the exception of jointed rocks). Furthermore, their behaviour is, once again, very sensitive to confining pressure that affects both the deformation characteristics and the failure mechanism.

The notion of pressure sensitivity of rocks is addressed in Figure 8.6. The figure shows the results of a series of drained 'triaxial' compression tests performed on porous limestone by Elliott and Brown [67]. The tests were carried out at various confining pressures ranging from 0.4 MPa to 30 MPa. It is evident from Figure 8.6a that at a low confinement the behaviour is predominantly elastic-brittle. The sample fails by formation of a macrocrack and the reduction in strength, after reaching the peak, is very abrupt. The brittle response is associated with a significant dilation, i.e. increase in volume triggered by a gradual opening of the macrocrack, Figure 8.6b. As the confining pressure increases, the behaviour becomes more ductile. In this case, the rate of strain softening decreases, the residual strength progressively increases and the rate of dilation is further reduced. At intermediate confining pressures, say between 5 MPa to 10 MPa,

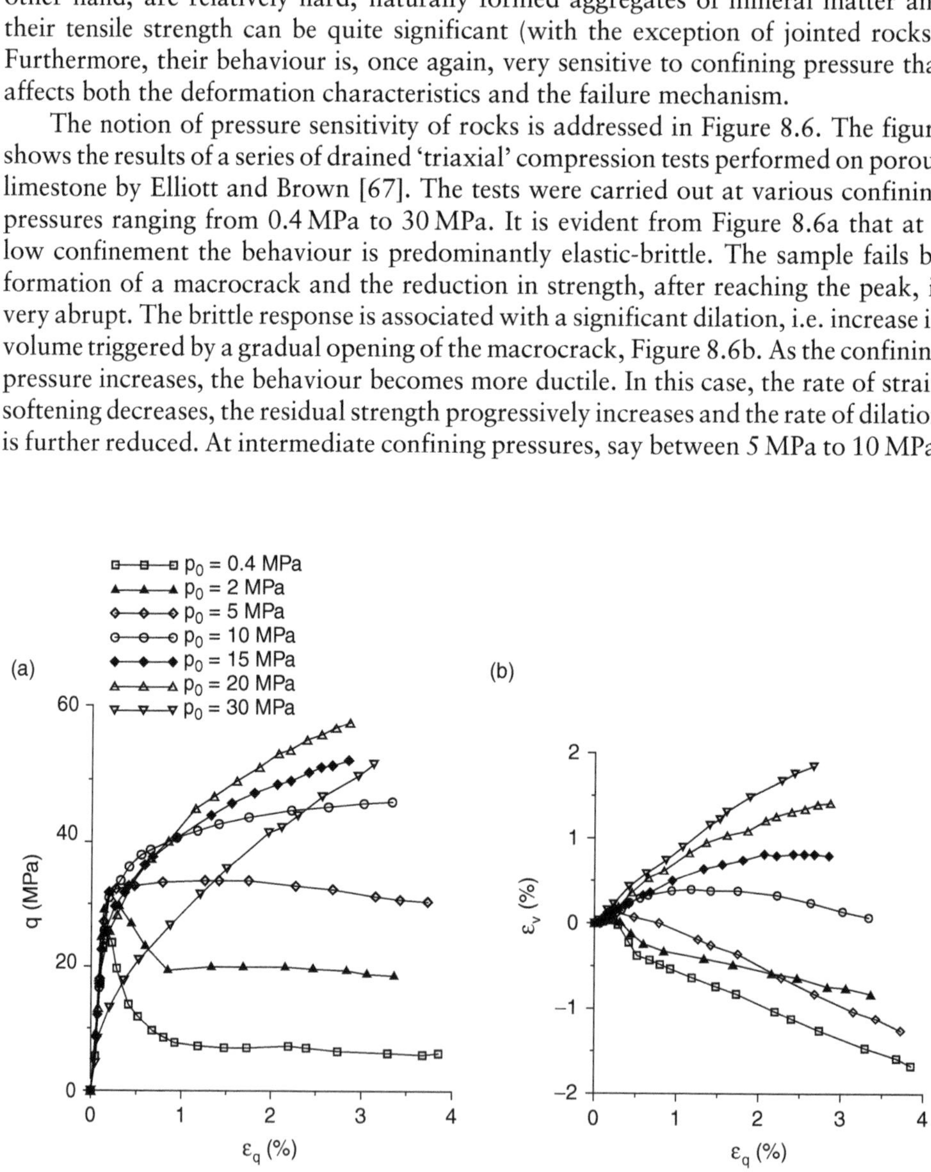

Figure 8.6 Axial compression tests on porous limestone at confinements range $p_0 = 0.4$–30 MPa (data from Ref. [67])

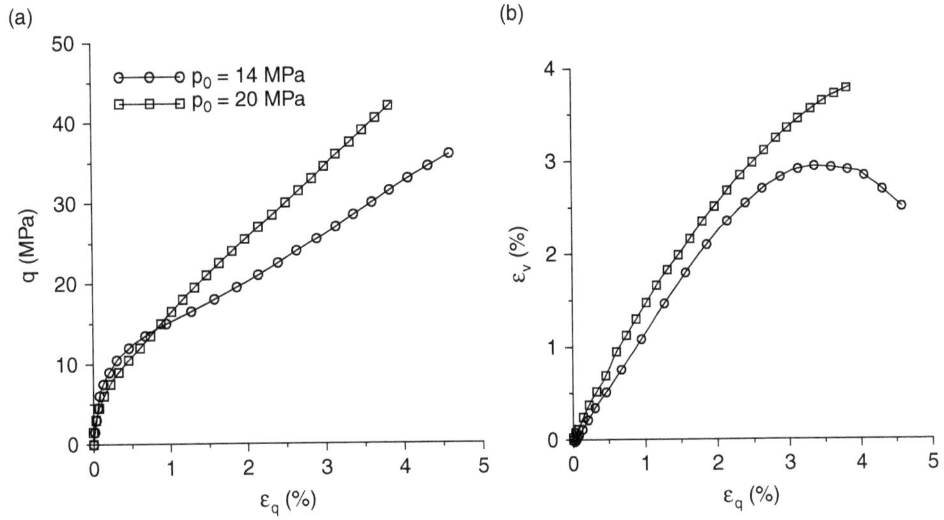

Figure 8.7 Results of axial compression on oil saturated samples of Lixhie chalk tested at high confinements (data from Ref. [68])

an unlimited plastic flow occurs at the peak stress level, so that the behaviour is characteristic of an elastic perfectly plastic material. The rate of volume change significantly decreases and the volumetric strain remains primarily compressive. Under high confining pressures (>15 MPa) the behaviour becomes typical of an elastic strain-hardening material. The ultimate load is reached at very high deviatoric strain (say >3%) and the material undergoes compaction, which progressively increases with an increase in confining pressure. The failure of the material typically involves the formation of compaction bands (i.e. narrow zones of localized purely compressive deformation), pore collapse and/or grain crashing. Under very high confining pressures (e.g., 30 MPa for the limestone examined here), the plastic deformation due to pore collapse occurs already in the first stage of the test, i.e. consolidation due the applied confinement. Therefore, at the beginning of the second stage, i.e. application of deviatoric loading, the axial deformation is significantly higher than that in tests with lower confining pressure, Figure 8.6a. In this case, no ultimate load is attained until a very high strain level. This kind of behaviour is also observed in other porous rocks, for example chalk (cf. Ref. [68]). This is illustrated in Figure 8.7, which shows results of 'triaxial' compression on oil saturated samples of Lixhe chalk tested at high confinements of 14 MPa and 20 MPa.

The 'triaxial' compression tests employed above, although useful within the context of qualitative/quantitative assessment of the material response, are restrictive as they involve an axisymmetric stress state, and thus, a fixed value of Lode's angle. In general, a complete characterization of the material behaviour requires a true triaxial configuration. This is of a particular importance for the specification of conditions at failure under a general stress state.

8.1.2 Influence of Lode's angle and the phenomenon of strain localization

Lode's angle defines the orientation of the stress vector in the octahedral plane (see eq.(2.13); Chapter 2). Its two extreme values correspond to conditions in standard 'triaxial' compression, $\theta=\pi/6$, and 'triaxial' extension, viz. $\theta=-\pi/6$. It is known that the material characteristics of porous media, in terms of strength and deformation response, are affected by Lode's angle. A typical example of this dependence is the asymmetry of the failure envelope in compression and extension domains of the $\{p,q\}$ space, as pointed out in Chapter 3 (cf. eqs.(3.8) and (3.10)). In general, in virtually all geomaterials (soils, rocks and concrete), the deviatoric stress at failure in axial/lateral compression is higher than that in axial/lateral extension. This is essentially due to the pressure sensitivity of these materials; in metals, for example, no such differences occur. The considerations of the influence of Lode's angle are of particular significance in constitutive modeling, as the general formulation should adequately address arbitrary loading histories.

Figure 8.8 shows the experimentally determined octahedral cross-sections of the failure surface for sandstone. The data was collected by Lade [69] and comes from a series of true triaxial as well as biaxial compression tests reported in Ref. [70]. The results correspond to confining pressure of approx. 83 MPa and invoke the assumption that the material is isotropic; i.e. the points have been projected into all six sectors of the octahedral plane. The cross-section is a curved triangular surface with smoothly rounded corners. Apparently, the maximum strength corresponds to 'triaxial' compression regime ($\theta=\pi/6$), while the minimum is obtained in 'triaxial' extension ($\theta=-\pi/6$).

Qualitatively similar results are obtained for soils. Figure 8.9 shows typical experimental data depicting failure envelopes in the octahedral plane for soils. The results in Figure 8.9a were again collected by Lade [69] and come from a series of tests on dense Monterey sand (initial void ratio $e_0 \simeq 0.57$) as reported in Ref. [71]. The cross-section shown corresponds to the confining pressure of 165 kPa. Figure 8.9b gives the results for loose Fuji river sand ($e_0 \simeq 0.84$) tested in a true triaxial apparatus at the confinement of 100 kPa (Ref. [72]). Note that for both loose and dense specimens the Mohr-Coulomb approximation, which is an irregular hexagon, appears to be sufficiently accurate, while the Drucker-Prager one (i.e. circular cross-section) is clearly inadequate.

For rocks, as well as cemented aggregate mixtures (such as concrete), the experimental evidence indicates that the shape of the octahedral section of the failure envelope undergoes evolution with the confining pressure. At low pressures, when the failure is brittle in nature, the cross-section is a curvilinear triangle, similar to that depicted in Figure 8.8. As the confinement increases however, there is a transition to ductile behaviour and the octahedral cross-section changes the curvature, becoming progressively more circular in shape (see, e.g. Ref. [69]).

The brittle failure in rocks/concrete is accompanied by formation of macrocracks. The associated mechanical characteristics become unstable in Drucker's sense (strain-softening). Similarly, in soils tested at relatively high degrees of initial compaction, the failure is characterized by the appearance of shear bands that is accompanied by strain-softening at the macroscale. The experimental evidence comes primarily from

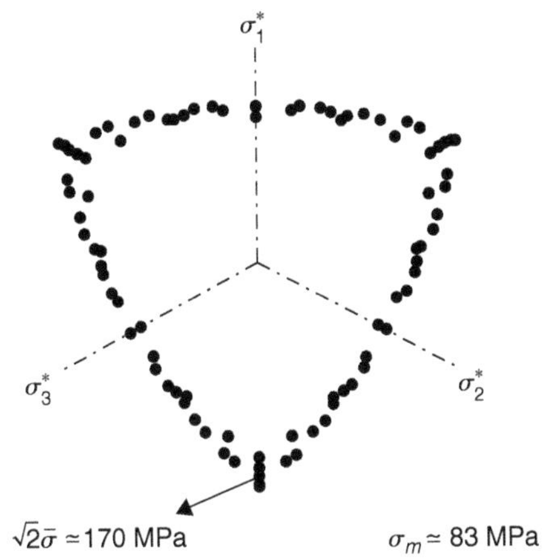

Figure 8.8 Octahedral section of the failure surface for sandstone (data from Ref. [69])

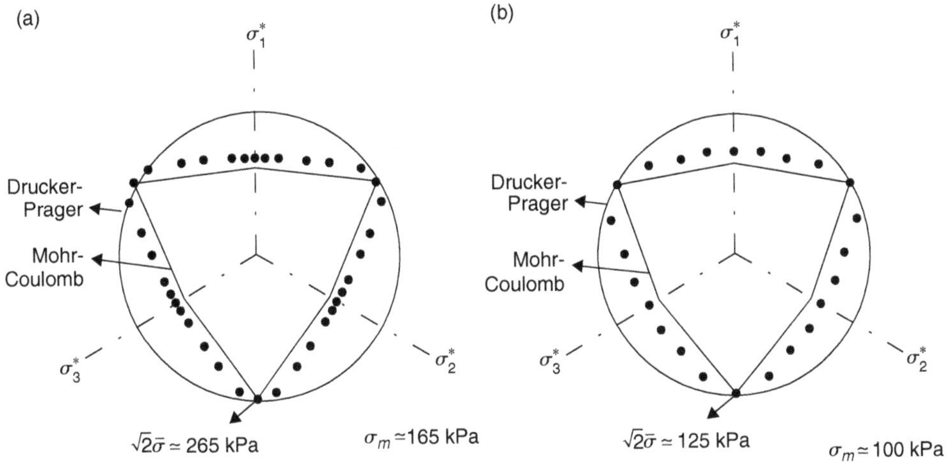

Figure 8.9 Failure envelopes in octahedral plane for: (a) dense Monterey sand (data from Ref. [69]), (b) loose Fuji river sand (data from Ref. [70])

true triaxial as well as biaxial compression tests. Figures 8.10 and 8.11 show typical data that depicts the localized deformation mode in both rocks and granular media. Figure 8.10 shows the results of a true triaxial test on Westerly granite, as reported by Haimson and Chang [73]. The test was carried out by applying a hydrostatic load until the desired value of the minor and intermediate stress was reached and then increasing the magnitude of σ_1 by maintaining a constant displacement rate along the minor principal stress direction. The thin cross-sections of the specimen were then

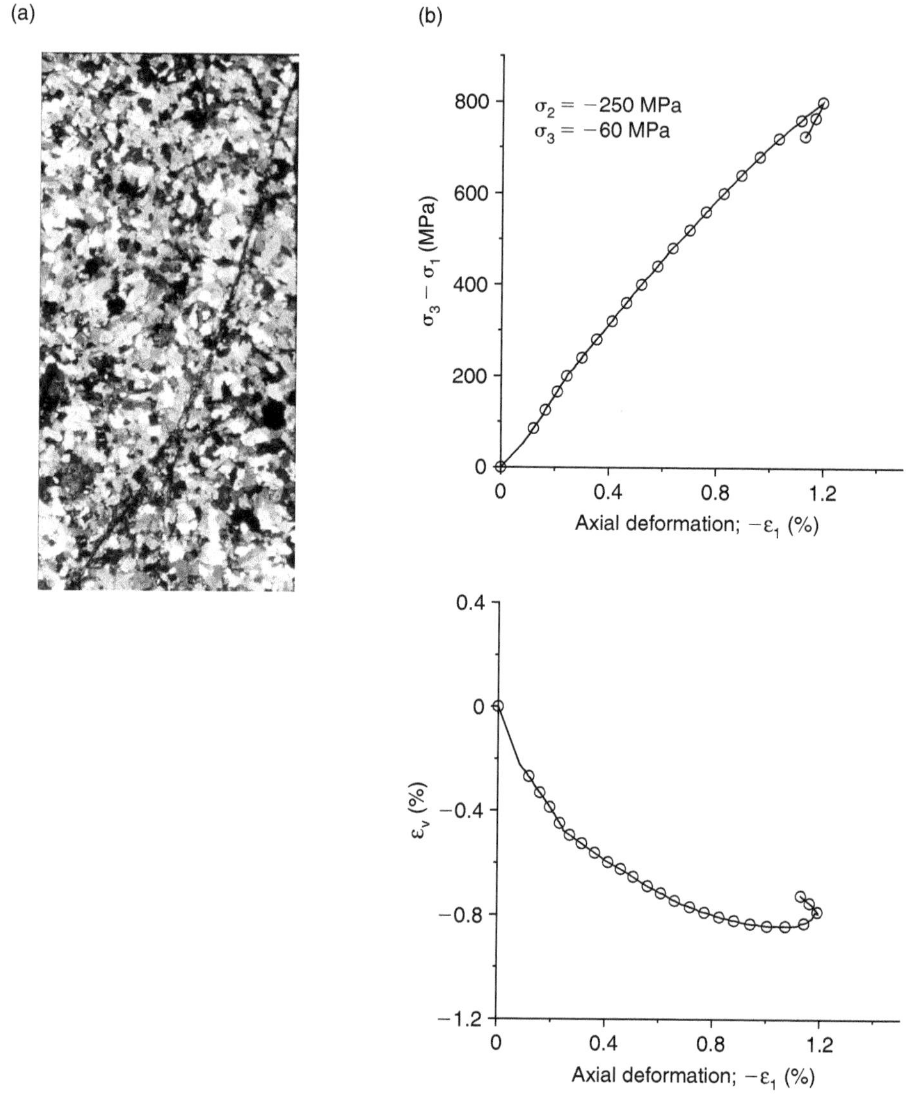

Figure 8.10 Results of a true triaxial test on Westerly granite (data from Ref. [73]); (a) cross-section of the failed specimen, (b) mechanical characteristics

scanned under the electron microscope. It is worth noting that, in this case, the post-peak instability is associated with a snap-back response, i.e. reduction in both axial stress and strain magnitudes. Figure 8.11 gives the results of a biaxial (plane strain) experiment conducted on a dense Ottawa sand ($D_r = 87\%$) at the confinement of 100 kPa (Khalid et al. [74]). Here, the specimen's deformation was recorded by digital monitoring of nodal displacements of the grid that was imprinted on the membrane surface (Figure 8.11a).

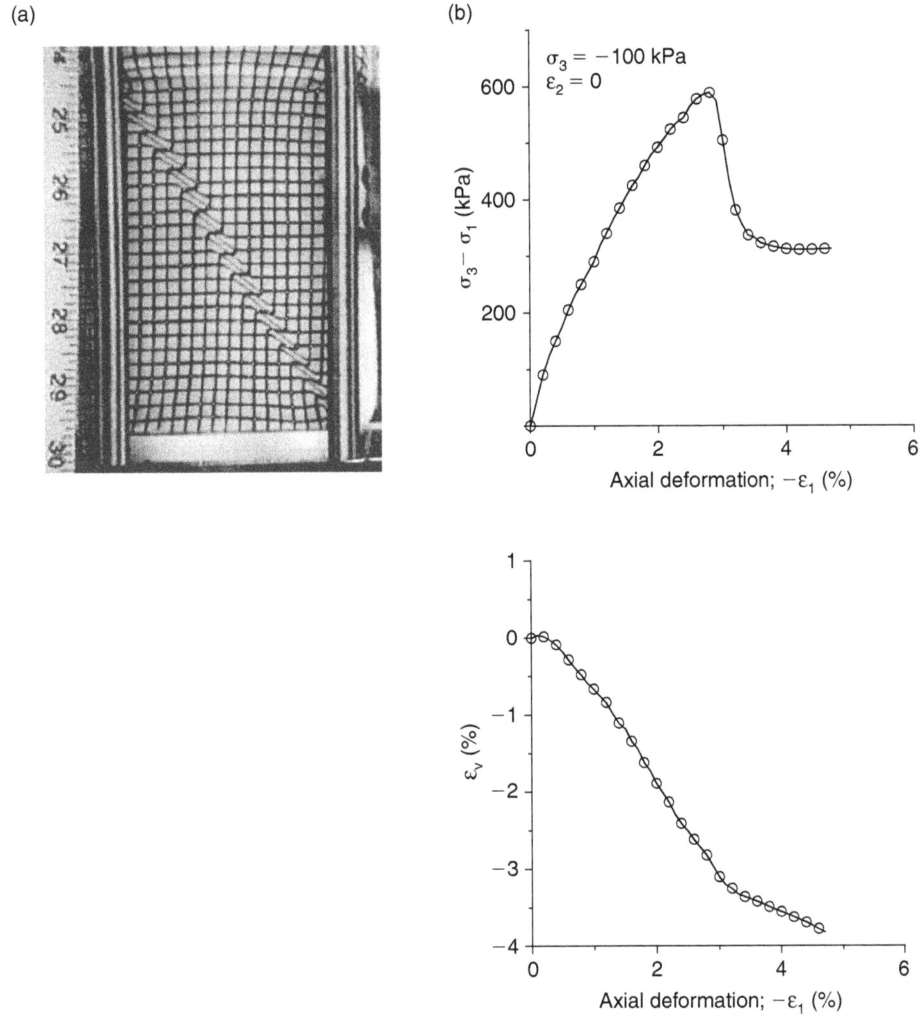

Figure 8.11 Results of a plane strain compression test on dense Ottawa sand (data from Ref. [74]); (a) failure mechanism, (b) mechanical characteristics

It is evident that in both cases considered above, the softening response is associated with an inhomogeneous deformation mode. It is therefore reasonable to argue that the mechanical characteristics recorded at the macroscale do not represent the response of the material per se, but that of a structure. The latter is a composite medium, whose mechanical behaviour is affected by the geometric aspects (viz. sample dimensions, orientation of macrocrack, etc.) and will also depend on the frictional properties along the macrocrack as well as the stiffness of the testing machine. In fact, if several initially homogeneous samples of the same material, with the same cross-sectional area but a different height, were tested under identical boundary conditions, the resulting strain-softening characteristics would significantly differ. Thus, if the strain-softening response is perceived as that of a classical continuum, the associate material parameters *cannot be uniquely defined.*

Another important implication of attributing strain-softening to the material behaviour, in a classical sense, is the fact that the finite element analysis of boundary value problems shows significant mesh size/alignment sensitivity. This has been first pointed out by Pietruszczak and Mroz [75]. To overcome the problem several conceptually different approaches have been developed that include non-local theories [76,77], micro-polar continua [78], higher-order gradient formulations [79,80], etc. The existing regularization procedures incorporate a characteristic dimension associated with strain-softening response that leads to mesh objectivity. An alternative to the fore mentioned approaches is a simple averaging scheme introduced in Ref. [75] and further refined in Refs. [81,82]. The latter is, in fact, the only methodology that accounts for anisotropy at the macroscale by distinguishing between the properties of the intact material and those along the macrocrack/shear band; both identifiable from standard experiments, i.e. 'triaxial' and direct shear tests.

Although none of the above frameworks are explicitly addressed in this book, it is important to stress that dealing with the strain-softening and the localized deformation requires advanced methodologies that go beyond the standard plasticity procedures examined earlier in Chapters 2–4.

8.2 UNDRAINED RESPONSE OF GRANULAR MEDIA; PORE PRESSURE EVOLUTION, LIQUEFACTION

In this section, the mechanical behaviour of geomaterials under undrained conditions is examined. A general review of qualitative trends in the undrained response of granular materials was given earlier in Section 3.1.3 of Chapter 3. Here, the actual experimental results are provided, for both soils and rocks, and the basic features are highlighted.

In a typical undrained test, the specimen is fully saturated and the fluid cannot escape from it as the drainage valve is closed. As a result, the overall macroscopic deformation remains consistent with that of the constituents. For soils, the deformability of grains forming the soil skeleton can be neglected and fluid itself (typically water) may be considered as incompressible compared to compressibility of the skeleton. In this case, the kinematic constraint of undrained deformation results in no volume change at the macroscale. This is not the case though for most rocks, where the compressibility of solid phase is comparable to that of fluid and a significant volume change may occur.

The tests are typically performed by controlling the axial displacement. In the case of soils, since the volume is said to be constant, only the axial stress and the excess of pore pressure are monitored. The results are commonly presented in terms of effective stress trajectories, in $\{p, q\}$ space, and the corresponding deviatoric stress-strain characteristics. The effective stress measures are obtained by invoking Terzaghi's principle, viz. eq.(3.16); i.e. $\hat{p}=p+p_w$; $\hat{q}=q$, where $\hat{p}$ and p is the *total* and *effective* pressure, respectively. This principle is considered valid since the actual contact areas between grains, in most soils, are infinitesimal. The mechanical response is strongly affected by the initial degree of compaction. In general, the behaviour under undrained constraint is governed by the volume change characteristics of soil skeleton that are measured under drained conditions. If the material displays a tendency to compaction (e.g., loose sand), this leads to a progressive build up of the excess of pore pressure. On the other hand, in a dilating material (dense sand) a generation of negative excess of pore pressure takes place. Given the complexity of volume change characteristics in sand, the macroscopic behaviour of the sample can be quite diversified and range from a stable response associated with strength exceeding that under drained conditions, to a complete liquefaction of the sample. The basic trends in mechanical characteristics have been depicted in Figure 3.5 in Section 3.1.3. In general, in dense sand, the plastic dilation triggers a build up of negative excess of pore pressure. As a result, the effective stress path moves away from the stress space origin and gradually approaches the failure envelope. On the other hand, in very loose sand, a significant generation of positive excess of pore pressure takes place. At some stage, which is associated with relatively small deviatoric strain, a loss of stability occurs, i.e. the deviatoric stress intensity decreases under continuing deformation. At the end of the test, the effective pressure reduces to zero; the contact between grains is lost and the material liquefies. Apparently, the stress trajectories corresponding to loose/medium dense sand are intermediate between the two extreme cases.

It needs to be noted that in all undrained tests the effective stress path is governed by the kinematic constraint of no volume change. Such a trajectory is unique in compression/extension domain and independent of the total stress path. Thus, different loading histories, at a fixed initial confinement, will generate the same effective trajectory; the evolution of pore pressure will be different though, as $p_w = \hat{p} - p$.

Figure 8.12 shows the results of undrained 'triaxial' tests on Reid-Bedford sand with the relative density of $D_r = 76\%$. The tests were conducted at confining pressures of 70 kPa and 700 kPa and the results were reported in Ref. [83]. Figure 8.12a gives the effective stress paths, while Figures 8.12b and 8.12c show the deviatoric characteristics and the evolution of excess pore pressure, respectively. The behaviour is, in general, consistent with the comments made earlier. Since the material is relatively dense, a significant negative excess of pore pressure develops and the deviatoric stress attains a value that is much higher than the corresponding strength under drained conditions. The sample fails, most likely, by cavitation of the pore fluid. The onset of cavitation is associated with volume change, so that the kinematic constrained of undrained deformation is no longer enforced.

The results shown in Figure 8.13 come from the same source (i.e., Ref. [83]) and correspond to loose Reid-Bedford sand with relative density of $D_r = 20\%$. Here, a significant positive excess of pore pressure develops and the effective pressure progressively decreases. The undrained strength is approximately half of that under drained conditions.

Finally, the results presented in Figure 8.14 correspond to Banding sand tested in a very loose state of compaction, i.e. at $D_r = 13\%$ (Castro [84]). The test was conducted at the confinement of 400 kPa. Again, the characteristics display the general trends outlined earlier. During the test, a generation of significant positive pore pressures takes place and the effective stress path migrates towards the stress space origin. At some

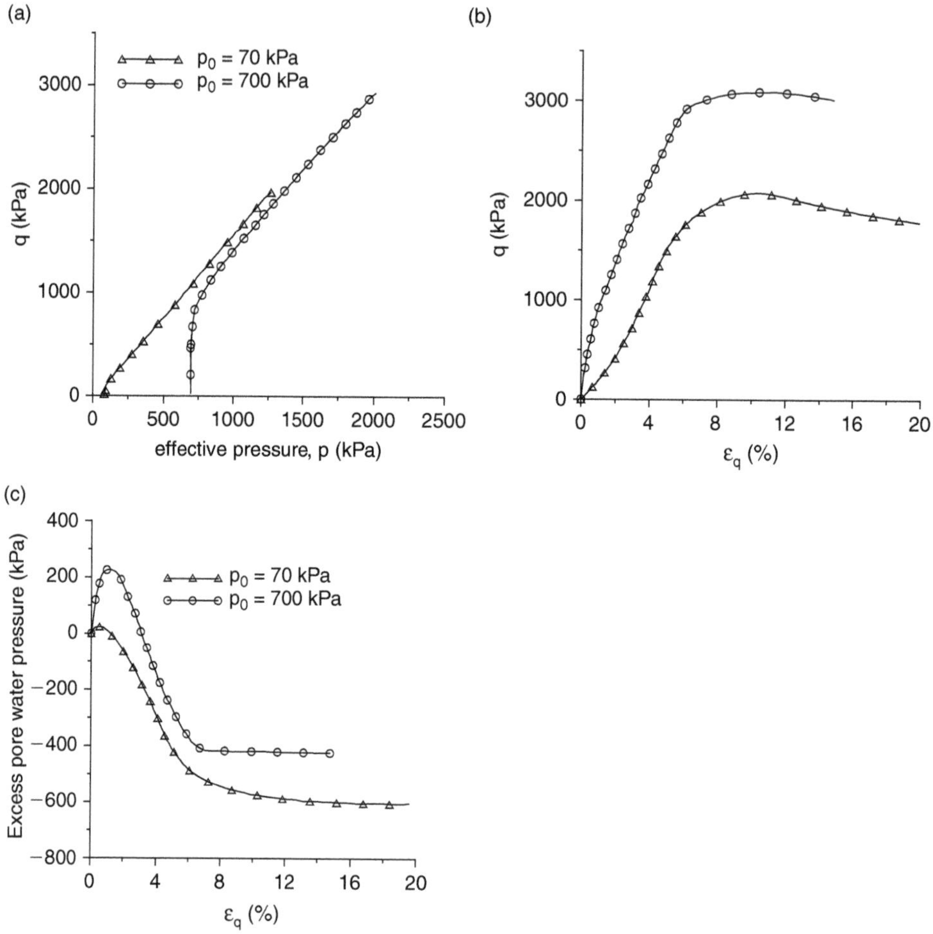

Figure 8.12 Results of undrained triaxial tests on dense Reid-Bedford sand (data from Ref. [83]); (a) effective stress trajectories, (b) deviatoric characteristics, (c) evolution of pore pressure

stage, the deviatoric characteristic becomes unstable, leading to a complete liquefaction of the sample. The latter manifests itself in the development of large deviatoric strain.

In rocks, the undrained deformation is associated with volume change and the Terzaghi's notion of effective stress does not, in general, apply. Therefore, during the test, the volumetric strain is monitored together with the evolution of interstitial pressure. The tests are primarily intended for the purpose of identification/verification of the coefficients in Biot's poroelastic or in poroplastic formulation (see, e.g. Ref. [85]). The mechanical response of rocks in undrained 'triaxial' tests strongly depends on the microstructure of the material and the confining pressure. In Figure 8.15 some typical results for sandstone are presented that involve compression tests, under undrained constraint, at the confining pressures of 7 MPa and 40 MPa (Sulem and

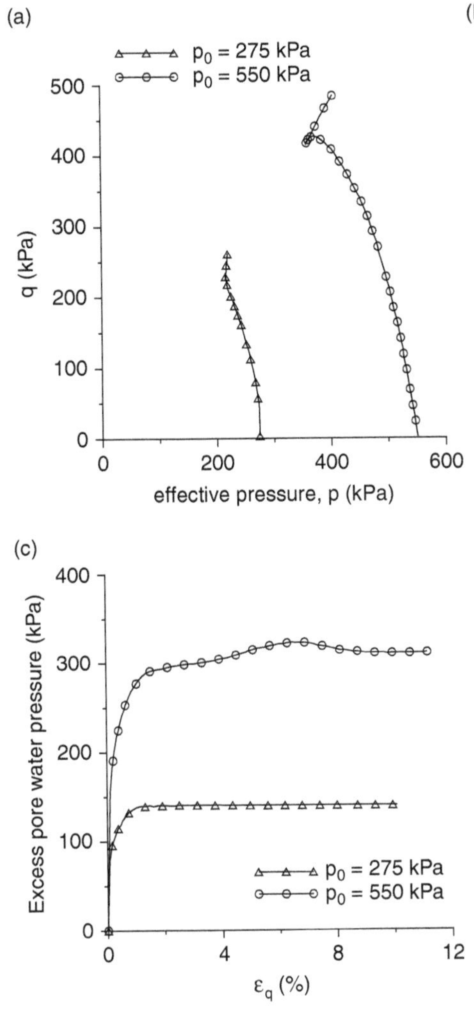

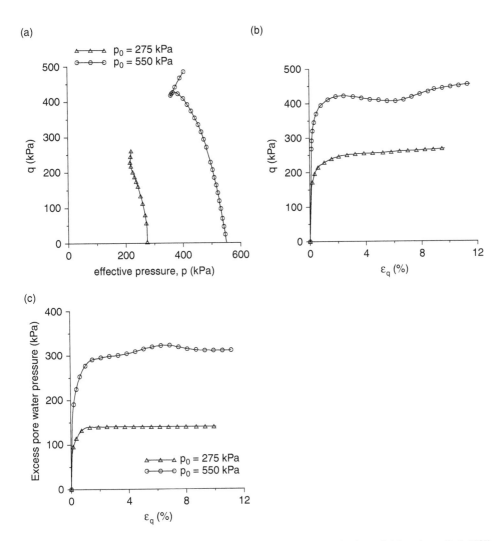

Figure 8.13 Results of undrained triaxial tests on loose Reid-Bedford sand (data from Ref. [83]); (a) effective stress trajectories, (b) deviatoric characteristics, (c) evolution of pore pressure

Ouffroukh [86]). At the early stages of the tests, Figure 8.15d, the samples undergo compaction and, consequently, a generation of positive excess of pore pressure takes place, Figure 8.15c. As the deformation advances, the material undergoes a progressive damage associated with the onset and growth of microcracks. This leads to volume increase (dilatancy) and thus, the build up of negative excess of pore pressure. The response is qualitatively similar to that of dense sand, except for the continuing volume change. The mechanical strength significantly increases with an increase in confining pressure, Figure 8.15b. Under very low confinements, the interstitial pressure may become negative that will trigger a capillary desaturation within the sample.

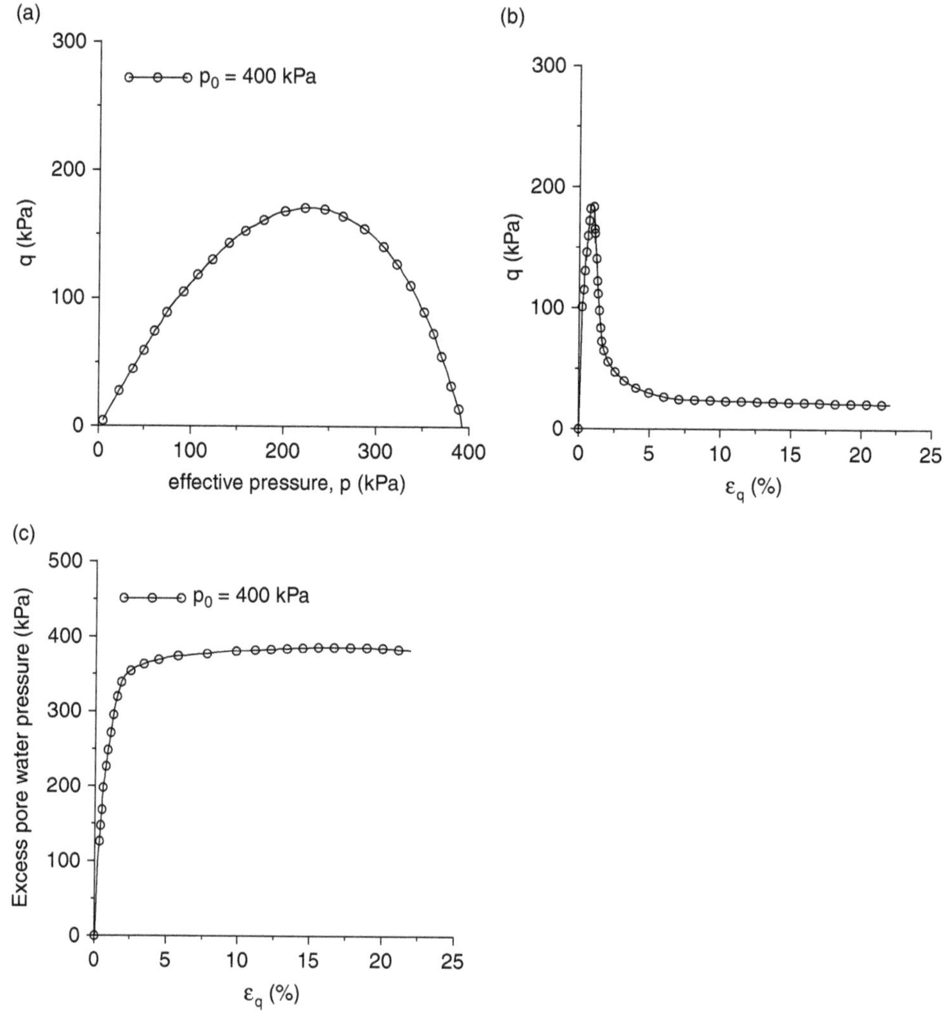

Figure 8.14 Results of an undrained test on a very loose specimen of Banding sand (data from Ref. [84]); (a) effective stress path, (b) deviatoric characteristic, (c) evolution of pore pressure

8.3 BASIC MECHANICAL CHARACTERISTICS IN CYCLIC TESTS; HYSTERESIS AND LIQUEFACTION

This section provides a brief outline of the basic trends in the mechanical behaviour of soils under cyclic loading. The focus is on granular media as their response is most diversified, especially within the context of undrained deformation. The interest in this topic is triggered by problems involving soil densification as well as seismic soil-structure interaction. Of particular interest is the study of devastating effects of liquefaction induced damage and the need for solutions that would mitigate its impact

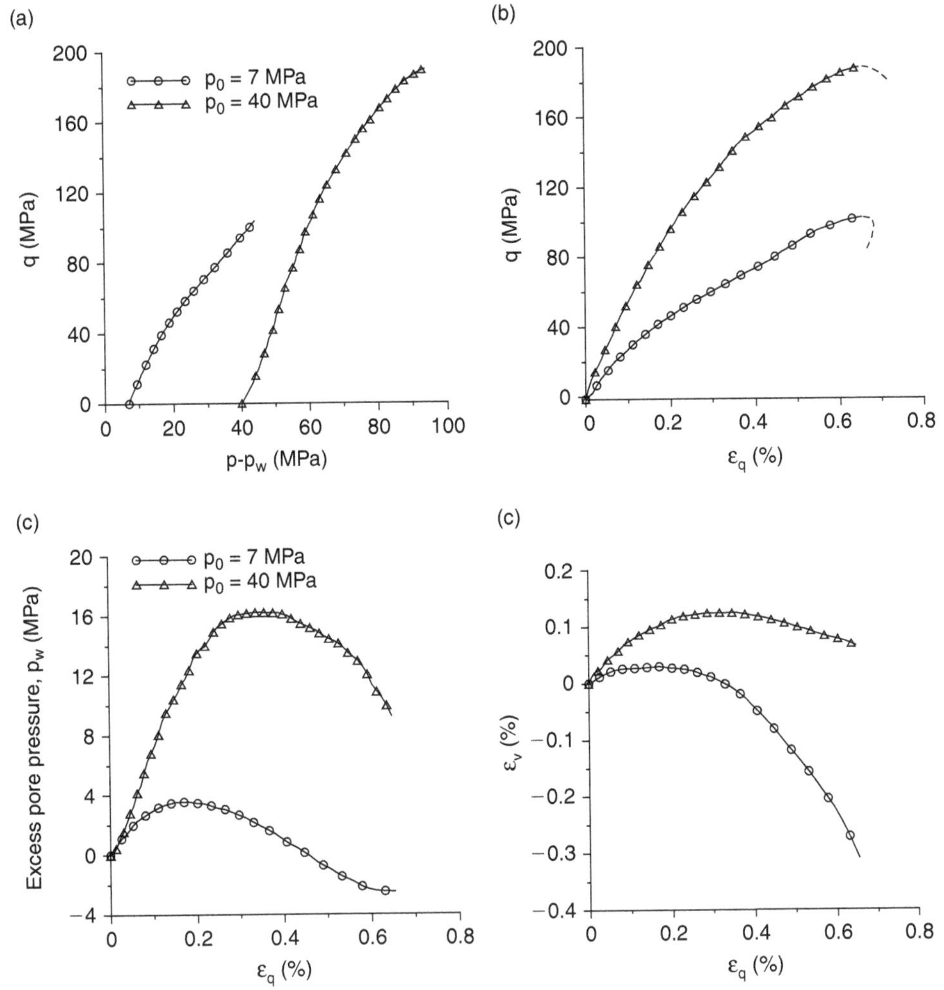

Figure 8.15 Results of undrained tests on sandstone (data from Ref. [85]); (a) stress trajectories, (b) deviatoric characteristics, (c) evolution of pore pressure and (d) volume change history

in future earthquakes. The analysis of this class of problems requires a comprehensive insight into both qualitative and quantitative aspects of the mechanical behaviour during consecutive load reversals.

Figure 8.16 shows a typical response of loose sand subjected to a number of *drained* 'triaxial' compression/extension cycles at fixed stress amplitude q/p. The tests were carried out on Fuji river sand ($e \simeq 0.75$) at confinement of $p_0 \simeq 200$ kPa (Tatsuoka and Ishihara [85]). Figure 8.16a shows the deviatoric characteristic, while Figure 8.16b gives the corresponding evolution of volume change for the case involving the stress amplitude of $q/p = \pm 0.8$. As the sample undergoes consecutive stress reversals, a progressive decrease in volume occurs. As the cycles continue, the rate of compaction

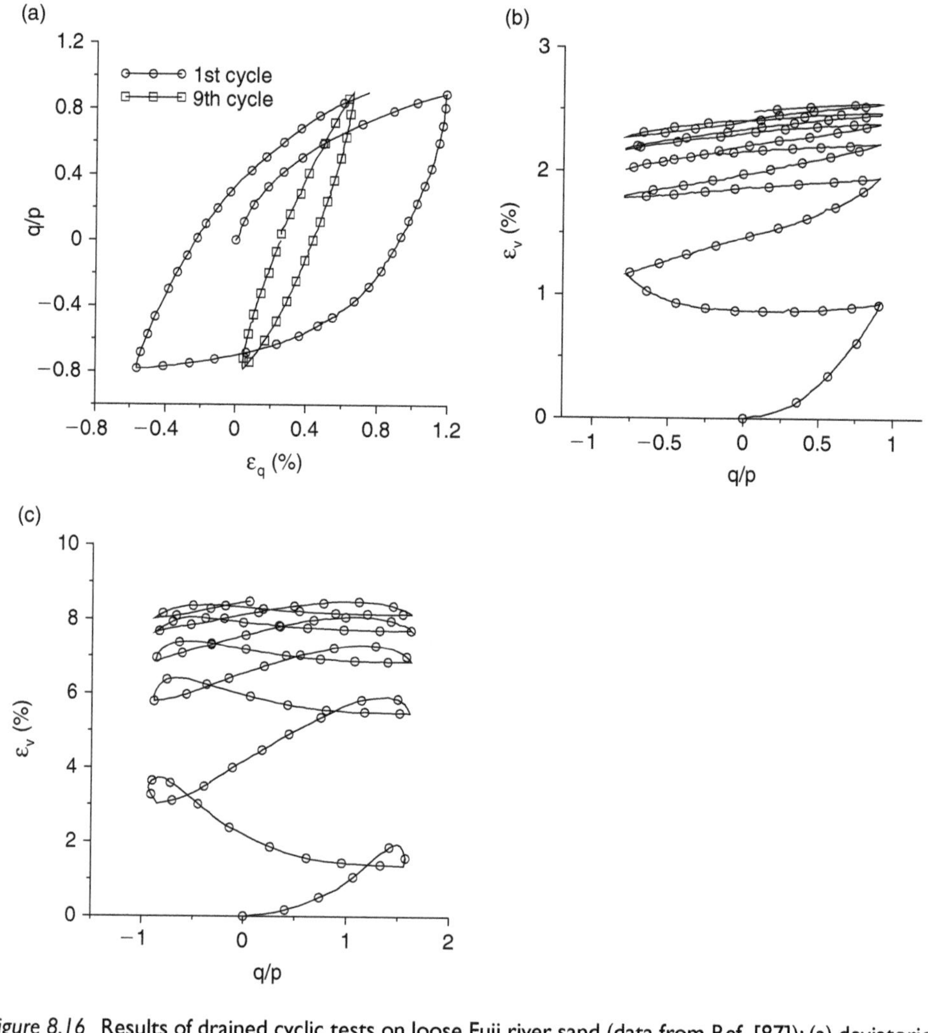

Figure 8.16 Results of drained cyclic tests on loose Fuji river sand (data from Ref. [87]); (a) deviatoric characteristic, (b) corresponding evolution of volume change, (c) evolution of volume change under larger stress-reversals

as well as the resulting deviatoric strain amplitude, both progressively decrease. The results shown in Figure 8.16c pertain to a sample of the same sand subjected to cycles with larger stress amplitudes in compression and extension. The trends are qualitatively similar to those depicted in Figure 8.16b. In this case, some degree of dilation occurs; however, the rate of volume change again progressively decreases as the cycles continue.

As mentioned earlier, if the soil skeleton displays a tendency to compaction, a progressive build up of excess of pore pressure develops under *undrained* conditions. The implications of this are evident in Figure 8.17, which shows the undrained response of loose/medium dense sand tested by Ishihara et al. [88]. The test was conducted

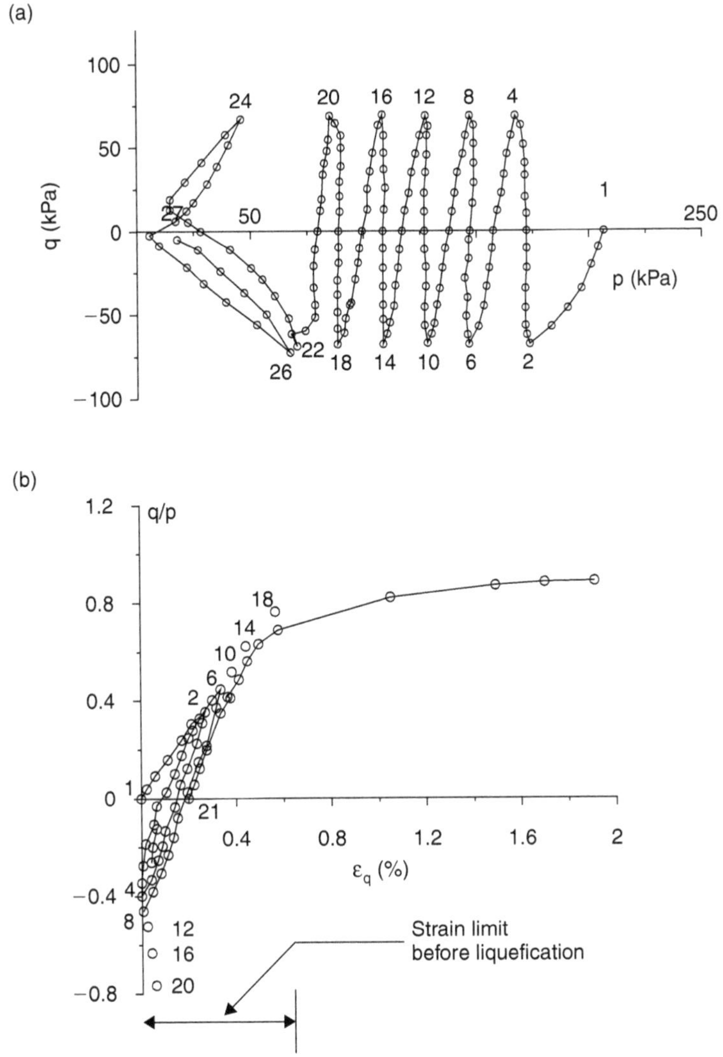

Figure 8.17 Results of an undrained stress-controlled cyclic test on loose Fuji river sand (data from Ref. [88]); (a) effective stress trajectory, (b) deviatoric characteristic

again on Fuji River sand (initial void ratio $e = 0.74$) at the confinement of 200 kPa and involved a fixed deviatoric stress amplitude of ±70 kPa. Clearly, during the test, the pore pressure progressively increases causing the effective stress trajectory to migrate towards the origin, Figure 8.17a. The positive excess of pore pressure is generated successively in both extension and compression domains and a gradual increase in the corresponding strain amplitudes is recorded. After a number of cycles the effective pressure abruptly reduces to zero; the contact between the grains is lost and the sample liquefies. The latter stage is associated with virtually unlimited deformation.

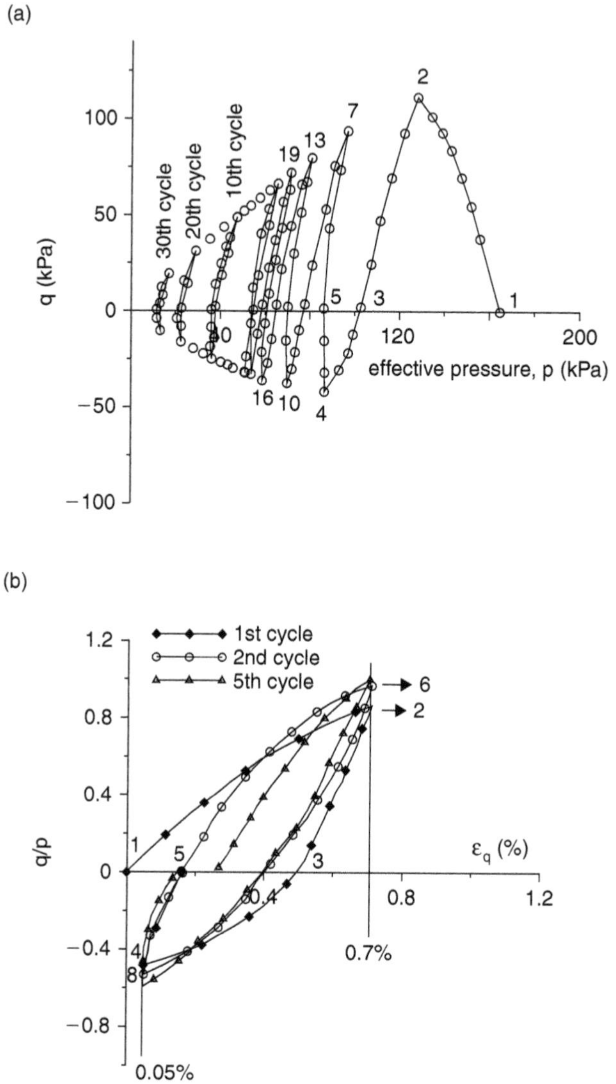

Figure 8.18 Results of an undrained strain-controlled cyclic test on loose Fuji river sand (data from Ref. [88]); (a) effective stress trajectory, (b) deviatoric characteristic

Figure 8.18 shows the response of the same material tested under strain-controlled regime, with deviatoric strain amplitudes fixed at 0.7% and 0.05%, respectively. Once again, during the test, the pore pressure progressively builds up causing the effective stress path to migrate towards the origin. This is associated with a gradual decrease in the corresponding stress amplitudes. After a number of cycles, the effective pressure drops to a small threshold value. However, since the strain amplitude is controlled here, no liquefaction per se is recorded.

8.4 INHERENT ANISOTROPY; STRENGTH CHARACTERISTICS OF SEDIMENTARY ROCKS

As pointed out in the preceding chapter, many geomaterials display inherent anisotropy, which is strongly linked with the microstructural arrangement. Such anisotropy may occur in granular media that comprise flat, elongated grains, but it is most typically associated with sedimentary rock formations, which are characterized by the existence of bedding planes. Examples include shale, schist, claystone, etc. In practical engineering applications, given the scale of the problem, these materials are commonly considered as continua with a transversely isotropic fabric. The inherent anisotropy manifests itself in the directional dependence of mechanical characteristics at the macroscale (Chapter 7).

The specification of the macroscopic response requires a series of tests that are conducted on inclined samples, i.e. extracted at different orientations relative to the bedding planes. The existing experimental evidence comes primarily from standard uniaxial/'triaxial' tests. Although, many quantitative aspects of these results are rather questionable (see Chapter 7), the tests still provide a valuable insight into the evolution of the failure mode. The results generally indicate that the maximum strength corresponds to a configuration in which the bedding planes are either parallel or perpendicular to the loading direction. At the same time, the minimum strength is typically associated with failure along the weakness planes.

Figure 8.19 shows typical results of a series of 'triaxial' compression tests carried out at different confining pressures and different orientation of the bedding planes relative to the loading direction. The data is taken from the article by Niandou et al. [89] and corresponds to Tournemire shale from the south of France. Figure 8.19a presents the variation of axial strength. The extreme values correspond to loading along one of the principal material axes. In general, the strength in the perpendicular direction is higher than that in the parallel one, which is related to the failure mechanism involved, Figure 8.19b. When the load is applied perpendicular to the bedding planes, the failure occurs by formation of microcracks in the matrix, followed by a localized mode involving the onset of macrocraking. When the load is parallel to the lamination, the failure also commences within the matrix but it is often associated with an abrupt decohesion of the bedding planes. The minimum strength is usually obtained for orientations ranging from 30° to 60° and the failure mechanism involves sliding along the bedding planes.

Figure 8.20 shows typical deformation characteristics. The results correspond to confining pressure of 40 MPa and involve different orientations of the bedding planes. It is evident that not only the strength, but also the deformation characteristics are significantly affected by the sample orientation. Here, the strength is the highest for the specimen tested at $\beta = 90°$ (vertical bedding planes). For all samples the behaviour is brittle after reaching the peak. In relation to volume change, Figure 8.20b, the sample with horizontal bedding planes undergoes a significant compaction. For the other orientations, viz. 45° and 90°, the volume change is less pronounced and at higher deviatoric stress intensities a transition from compaction to dilatancy occurs. Again, it should be mentioned that for inclined samples the deformation is no longer axisymmetric and the results should be taken with caution.

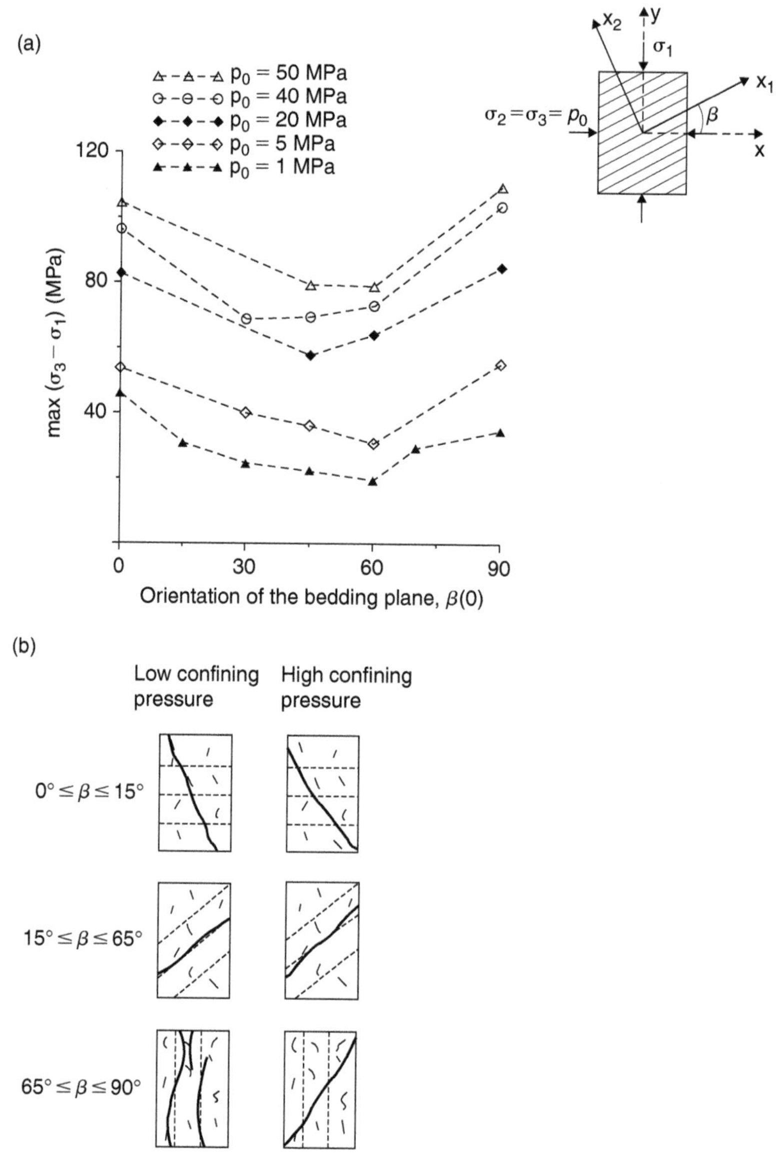

Figure 8.19 Results of triaxial tests on Tournemire shale at different orientations of bedding planes (data from Ref. [89]); (a) variation of axial strength at different confining pressures, (b) typical failure mechanisms

In fact, all specimens tested at orientations other than 0° and 90° will have the tendency to distort under the increasing axial load. Since in a 'triaxial' cell such distortion is kinematically constrained, this will affect the stress field within the sample.

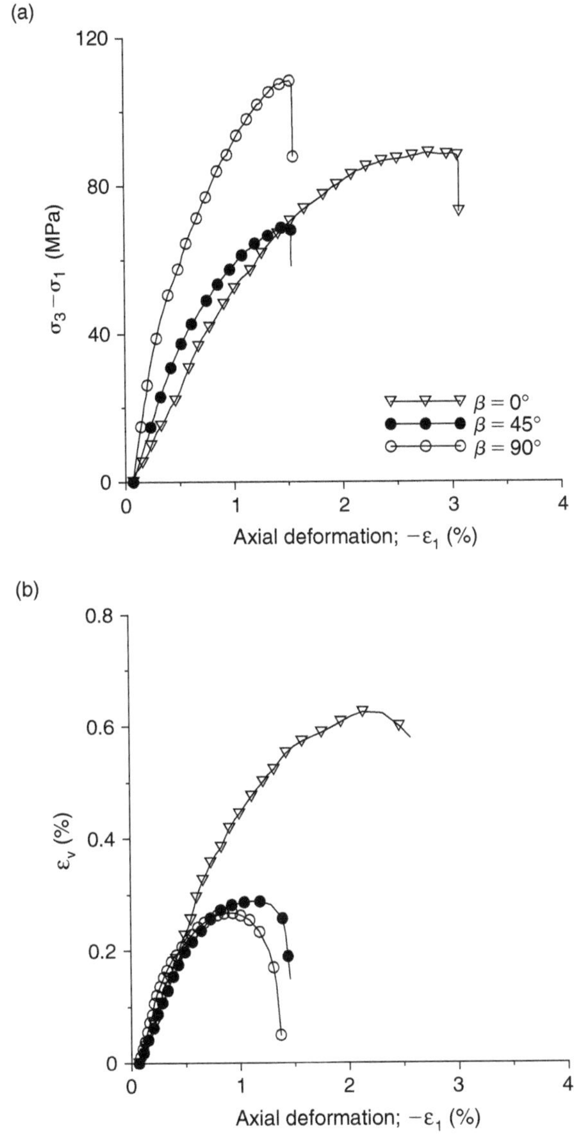

Figure 8.20 Results of triaxial tests on Tournemire shale at different orientations of bedding planes (data from Ref. [89]); (a) deviatoric and (b) volume change characteristics at confinement of 40 MPa

8.5 IDENTIFICATION OF BASIC MATERIAL PARAMETERS FOR SOILS/ROCKS

In this section, the procedures are outlined for specification of constants that are embedded in constitutive relations discussed earlier in Chapter 3. The focus is on granular materials as well as rocks that have a homogeneous random microstructure. Typical methodologies are reviewed, which employ conventional 'triaxial' tests for identification of parameters defining both the conditions at failure and the deformation response. Later, examples are given for specific materials; in particular, dense and loose Karlsruhe sand and limestone whose characteristics were examined in Section 8.1.1 of this chapter.

8.5.1 General remarks on identification procedures

The strain-hardening frameworks described in Chapter 3 were formulated first in the 'triaxial' $\{p, q\}$ space and later generalized for arbitrary stress trajectories. Such a methodology implies that the identification of basic material constants can be carried out using a series of conventional 'triaxial' tests, and the performance can later be verified for more complex loading configurations. For most soils, the conditions at failure are defined using Mohr-Coulomb representation, eq.(3.8), which employs two independent material parameters that are a function of the angle of internal friction and cohesion, eq.(3.9). Thus, for this class of materials, a minimum of two tests are required for the identification purpose; a better statistical representation though can be obtained by conducting a larger number of experiments. Clearly, any 'triaxial' trajectory that brings the specimen to a failure is admissible; including tests under drained/undrained conditions in both compression and extension domains. Typically, a series of conventional drained axial compression tests, performed at different confinements, is employed as illustrated in the example given in Figure 8.3. A similar methodology is followed in the context of rocks that remain isotropic at the macroscale. Here, the conditions at failure can be approximated using a quadratic form, viz. eq.(2.45) discussed in Chapter 2.

The deformation characteristics are described by employing an associated/non-associated flow rule, together with the notion of strain-hardening that can be attributed to either the irreversible volume change or plastic distortion. The question of whether the flow rule is actually associated or non-associated, can apparently be answered only within the context of a specific mathematical framework. In general, the approaches based on volumetric or combined volumetric-deviatoric hardening employ an associated flow rule; their range of applicability though is fairly restrictive, as discussed in Chapter 3. On the other hand, the framework based on deviatoric hardening, Section 3.3, incorporates a non-associated flow rule that is required in order to account for a progressive transition from compaction to dilatancy. The information on the material parameter that governs this transition can be obtained from the same set of conventional 'triaxial' tests as before, by examining the nature of volume change characteristics. An example, in the context of both granular media and rocks, is given later in this section.

It is noted that in porous materials, the direction of plastic flow may, in general, depend on the direction of the stress rate. There is an evidence of this in, for example, granular materials (cf. Ref. [90]). Clearly, this dependency is contradictory to the basic notion of the potential flow theory and cannot be accounted for in the standard plasticity framework. In fact, only the approaches that incorporate Koiter's flow rule [91], associated with existence of multiple loading surfaces, can include this effect.

Finally, the hardening characteristics employ one or more constants (depending on the complexity of the formulation) that can also be defined from conventional 'triaxial' tests. The frameworks of volumetric or volumetric-deviatoric hardening require a hydrostatic compression test, while in the approach incorporating deviatoric hardening the same set of axial compression tests, as discussed before, can be used. In what follows, examples are given of identification procedures for assessment of parameters embedded in the deviatoric hardening framework. Both granular soils and limestone are considered and the identification procedure employs the experimental results provided in Section 8.1.1.

8.5.2 Examples involving deviatoric hardening framework

For soils, the formulation of the problem involves the loading and plastic potential surfaces in the general form consistent with eq.(3.67), i.e.

$$f = \sqrt{3}\bar{\sigma} - \eta\sigma_m g(\theta) - \mu = 0; \quad \eta = \eta_f \varepsilon_q^p/(A + \varepsilon_q^p);$$

$$\psi = \sqrt{3}\bar{\sigma} + \eta_c g(\theta)\sigma_m \ln\frac{\sigma_m}{\sigma_m^0} = 0$$

which for the 'triaxial' configuration simplifies to

$$f = q - \eta p - \mu = 0; \quad \eta = \eta_f \varepsilon_q^p/(A + \varepsilon_q^p); \quad \psi = q + \eta_c p \ln\left(\frac{p}{\bar{p}}\right) = 0$$

In order to illustrate the procedure for identification of material parameters, consider the Karlsruhe sand, for which the results of drained 'triaxial' tests are provided in Figures 8.1–8.2. The issue of the specification of the conditions at failure has already been addressed in Section 8.1.1. Figure 8.3 shows the failure envelopes, in $\{p, q\}$ space, for both loose and dense states of compaction. Based on those results, the slope of the best fit approximation corresponding to Mohr-Coulomb criterion is $\eta_f = 1.26$ and $\eta_f = 1.61$ for loose and dense sand, respectively, while $\mu = 0$; i.e., the material poses no resistance to tension.

The specification of hardening function involves the identification of material constant A. In order to accomplish that, the mechanical characteristics are now re-plotted in $\{\varepsilon_q^p, \eta = q/p\}$ space. Note that constructing such characteristics requires the identification of the shear modulus G, so that the plastic deviatoric strain can be determined via the additivity postulate. The shear modulus should normally be assessed from the initial slope of the *unloading* branch, i.e. $3G = dq/d\varepsilon_q^e$. It should be stressed that its value may, in general, depend on both the confining pressure as well as the deviatoric stress intensity. Here, the experimental information is limited to active loading histories only. Given this restriction, the value of G was estimated from the initial slope of the *loading* branch. In order to maintain simplicity, a constant value of $G = 45$ MPa was selected which is representative of that at high confining pressures.

Figure 8.21a presents the hardening characteristics for loose Karlsruhe sand. At the same time, Figure 8.21b shows the best fit approximation employing the hyperbolic representation viz. eq.(3.62). It is evident that these characteristics are not significantly sensitive to the value of confining pressure, so that the simple framework employed here is quite adequate. The value of the parameter A is estimated here to be $A = 0.0059$.

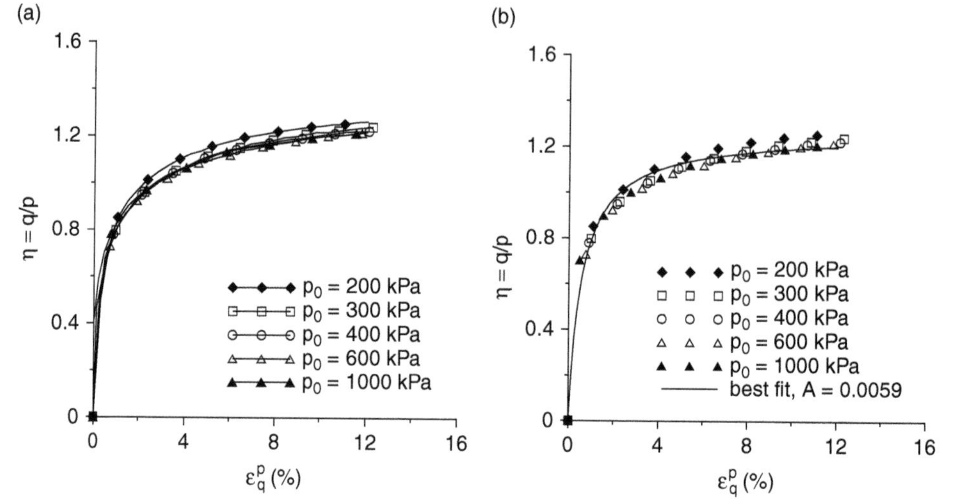

Figure 8.21 Hardening characteristics for loose Karlsruhe sand (data from Ref. [65]); (a) experimental response, (b) best fit approximation

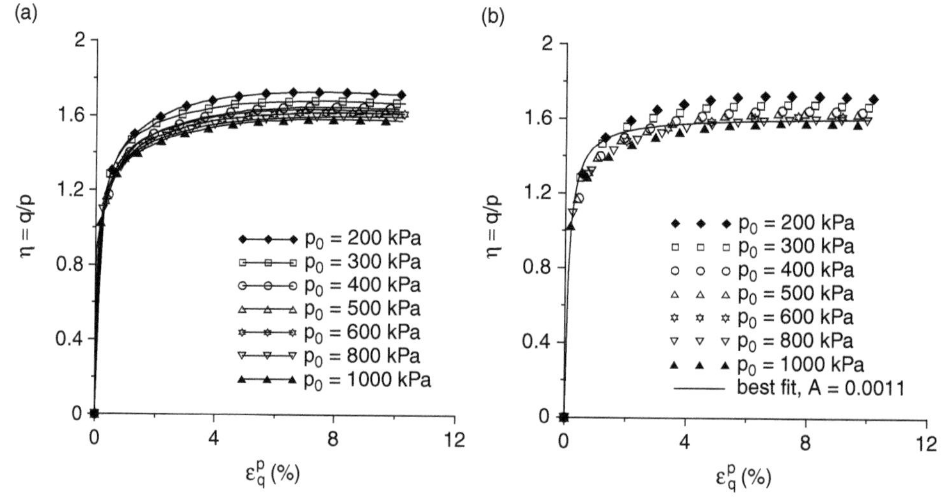

Figure 8.22 Hardening characteristics for dense Karlsruhe sand (data from Ref. [65]); (a) experimental response, (b) best fit approximation

Figure 8.22 shows similar results for dense Karlsruhe sand. The hardening characteristics were plotted here for G = 120 MPa and again cover a broad range of confinements. Figure 8.22b gives the best fit approximation that corresponds to $A = 0.0011$.

The next step of the identification procedure involves the specification of the plastic potential function, viz. the parameter η_c. The latter defines the transition from plastic

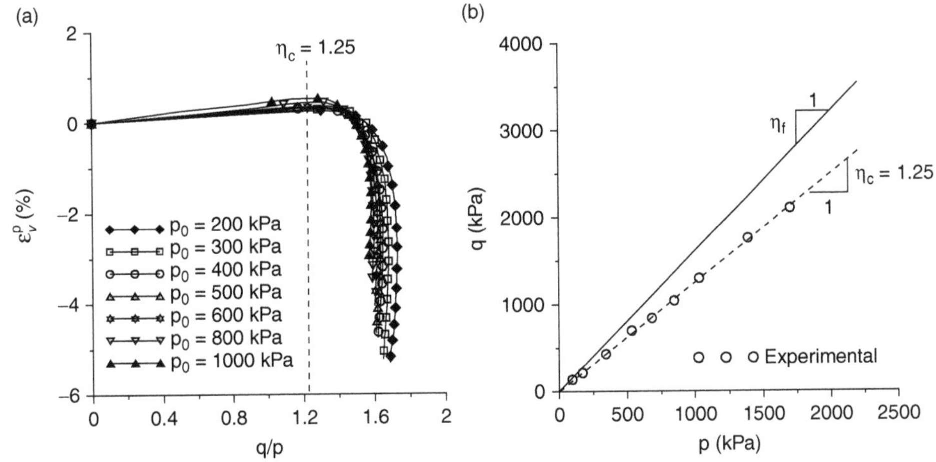

Figure 8.23 Specification of the onset of dilation for dense Karlsruhe sand; (a) evolution of volume change, (b) zero-dilatancy envelope

compaction to dilatancy and its value can be assessed by examining the respective volume change characteristics. Note that $d\varepsilon_v^p = 0$ requires $\partial\psi/\partial p = 0$, so that the locus of transition points is defined as $\eta = \eta_c$ or $q - \eta_c p = 0$. It is evident from Figure 8.1b that in case of loose sand, no dilation takes place and the ultimate state is associated with no volume change. This is consistent with selecting $\eta_c = \eta_f$. For dense sand, the volume change characteristics of Figure 8.2b need to be re-plotted in the affined space $\{q/p, \varepsilon_v^p\}$, Figure 8.23. The irreversible volume change is estimated here by assuming that $K = 3G$ and invoking the additivity postulate. Note that, in general, a sufficient degree of accuracy may be maintained by associating the transition points with the evolution of total volume. Figure 8.23b shows the zero-dilatancy line obtained by a linear best fit; the experimental points correspond to the maxima of the respective volume evolution curves. The estimated value of η_c is approximately 1.25, so that $\eta_c = 0.78\,\eta_f$.

Based on the considerations above, a set of material parameters can now be chosen for both loose and dense Karlsruhe sand, and the performance of the framework can be verified for various loading histories. Figure 8.24 shows the numerical simulations of a series of axial compression tests corresponding to loose Karlsruhe sand. The simulations were carried out assuming

$$G = 45\,\text{MPa}, \quad K = 3G, \quad \eta_f = \eta_c = 1.26, \quad A = 0.0059$$

and the results, for confining pressures ranging from 300 kPa to 1000 kPa, are compared with the experimental data of Figure 8.1. The predictions, in terms of both deviatoric characteristics and the evolution of volume change, appear to be fairly reasonable. Similar results, corresponding to dense sand, are shown in Figure 8.25. The material parameters selected here are

$$G = 120\,\text{MPa}, \quad K = 3G, \quad \eta_f = 1.61, \quad \eta_c = 1.25, \quad A = 0.0011$$

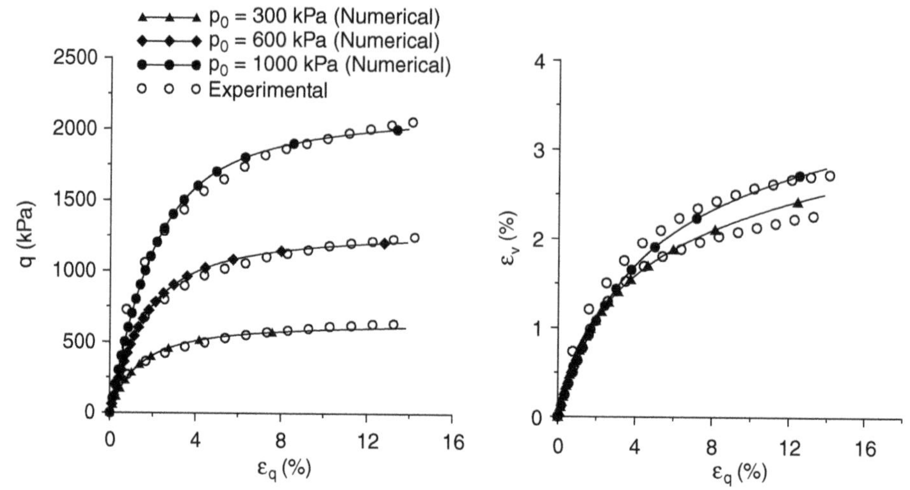

Figure 8.24 Numerical simulations of triaxial tests on loose Karlsruhe sand

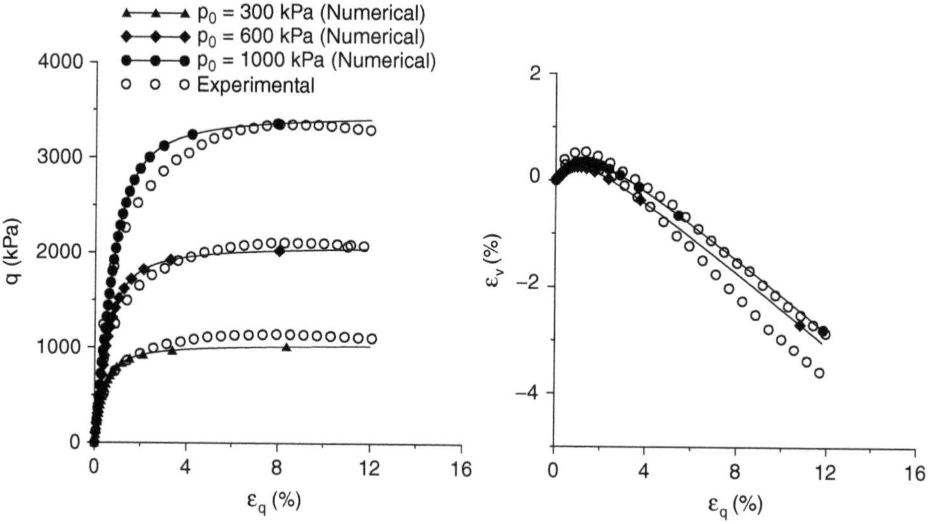

Figure 8.25 Numerical simulations of triaxial tests on dense Karlsruhe sand

and the simulations cover a similar range of confining pressures as before. Again, the basic trends are consistent with experimental results shown in Figure 8.2.

Consider now the mechanical characteristics of limestone, as depicted in Figure 8.6. In this case, the description of conditions at failure requires a quadratic form, viz. eq.(2.47). Thus,

$$F = a_1\left(\frac{\bar{\sigma}}{f_c\, g(\theta)}\right) + a_2\left(\frac{\bar{\sigma}}{f_c\, g(\theta)}\right)^2 - \left(a_3 + \frac{\sigma_m}{f_c}\right) = 0$$

where a's are dimensionless parameters and f_c is, in general, an arbitrary constant used for normalization purposes. Note that if f_c is identified with the uniaxial compressive strength then the number of independent parameters can be formally reduced to two. In the 'triaxial' configuration, the representation above can be simplified to

$$F = c_1 q + c_2 q^2 - (p + c_3) = 0 \quad or \quad F = q - q_c = 0,$$

$$q_c = \frac{-c_1 + \sqrt{c_1^2 + 4c_2(p + c_3)}}{2c_2}$$

where $c_1 = a_1/\sqrt{3}, c_2 = a_2/(3f_c), c_3 = a_3 f_c$. The yield/loading and the plastic potential surfaces can now be defined as

$$f = q - \beta q_c = 0; \quad \beta = \beta_0 + (1 - \beta_0)\varepsilon_q^p/(A + \varepsilon_q^p);$$

$$\psi = q + \eta_c(p + c_3)\ln\left(\frac{p + c_3}{\bar{p}}\right) = 0$$

Here, $\beta_0 \leq \beta \leq 1$ where β_0 is a threshold value that defines the initial yield surface.

The above formulation is a simple extension of the deviatoric hardening framework that can be applied to isotropic rock formations in the compression regime. It needs to be emphasized that this specific formulation is quite restrictive. In particular, it does not address the issue of localization of deformation that typically occurs at low confining pressures. A broader discussion on this topic is provided in Ref. [81].

Let us focus now on the identification of material parameters. Again, the issue of specification of conditions at failure needs to be addressed first. Based on the experimental data selected from Figure 8.6, which include initial pressures in the range from 0.4 MPa to 20 MPa, the best fit approximation has been obtained that corresponds to the quadratic form employed here (Figure 8.26). Assigning now to f_c a unit value (in units of pressure), the respective dimensionless strength parameters are

$$a_1 = 0.639, \quad a_2 = 0.0174, \quad a_3 = 3.90$$

The hardening characteristics have been defined following a procedure similar to that used for soils. The parameters β_0 and A have been identified by providing the best fit approximation to the experimental data plotted in $\{\varepsilon_q^p, q/q_c\}$ space. Figure 8.27 shows the experimental characteristics covering the range of confinements from 10 MPa to 20 MPa, together with the best fit curve; both corresponding to a constant value of G = 5.5 GPa. It needs to be emphasized that, in this case, the hardening characteristics are significantly affected by the confining pressure. The estimated values of $\beta_0 = 0.38$ and $A = 0.003$ are thus the averages that are representative over *a priori* selected range of pressures $p_0 \in [10, 20]$ MPa.

The plastic potential has the functional form similar to that employed for sand. The specification of the parameter η_c is based again on evaluation of the stress ratio at the onset of plastic dilation (cf. Figure 8.23b) under the assumption that the bulk modulus is equal to $K = 2G$. By examining the volume change characteristics in

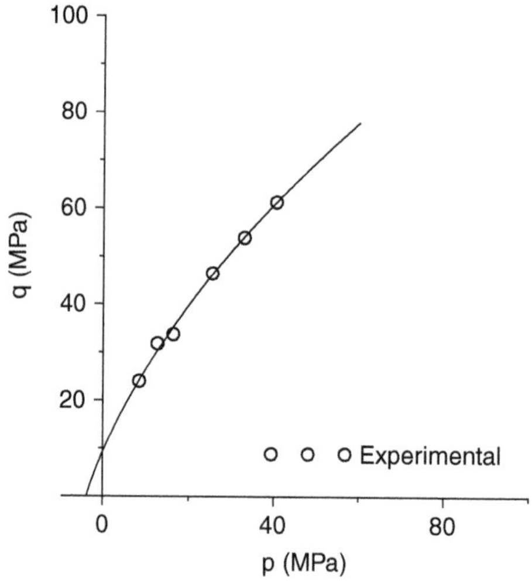

Figure 8.26 Meridional section of the failure envelope for limestone

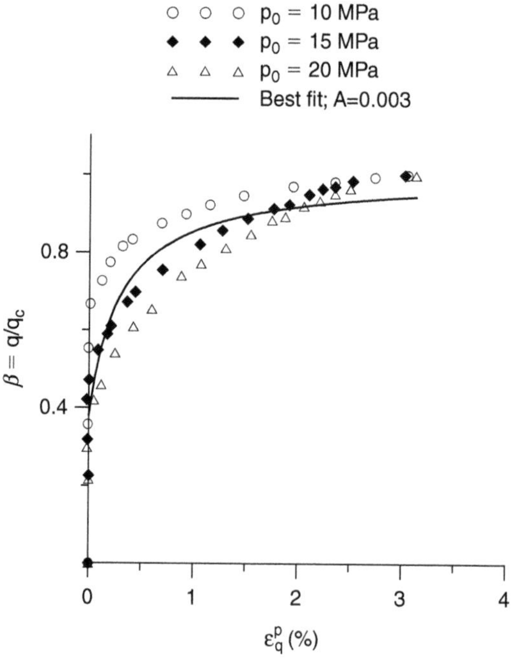

Figure 8.27 Hardening characteristics of limestone in the range of confinements from 10 MPa to 20 MPa (data from Ref. [67])

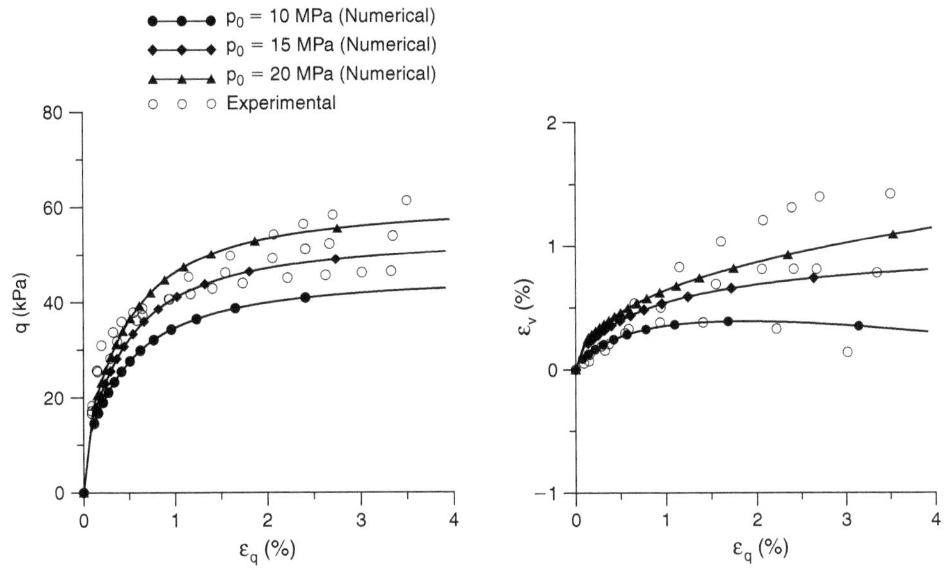

Figure 8.28 Numerical simulations of triaxial tests on limestone

Figure 8.6 over the relevant range of confinements, and employing a linear approximation to the zero-dilatancy envelope, the value of η_c was estimated as 1.45.

Figure 8.28 shows the numerical simulations of a series of axial compression tests. The material parameters correspond to the values identified above, viz.

$$G = 5.5\,\text{GPa}, \quad K = 2G, \quad \beta_0 = 0.38, \quad A = 0.003, \quad \eta_c = 1.45$$

The simulations cover the initial pressures of $p_0 = 10\,\text{MPa}$, 15 MPa and 20MPa. The response of the material is ductile, which is consistent with the experimental data for this range of confinements. Given the overall simplicity of the framework, the predictions in terms of deviatoric characteristics as well as the evolution of volume change seem to be quite reasonable. A more general description that incorporates both the pressure dependency of hardening function as well as localized deformation mode, and thus extends the applicability of this framework to a broader range of confinements, is given in Ref. [81].

Bibliography

[1] Hill R. *The mathematical theory of plasticity*. Clarendon Press, Oxford (1950).

[2] Prager W. *An introduction to plasticity*. Addison-Wesley, Reading, MA (1959).

[3] Chen WF, Han DJ. *Plasticity for structural engineers*. Springer-Verlag, New York (1988).

[4] Chen WF, Baladi GY. *Soil plasticity, theory and implementation*. Elsevier Science, Amsterdam (1985).

[5] Davis RO, Selvadurai APS. *Plasticity and geomechanics*. Cambridge University Press, Cambridge (2002).

[6] Smith GF. *Constitutive equations for anisotropic and isotropic materials*. North-Holland, Amsterdam (1994).

[7] Boehler JP (Ed). *Application of tensor functions in solid mechanics*. CISM Courses and Lecture Notes, No. 292. Springer-Verlag, New York (1987).

[8] Prager W. A new method of analyzing stresses and strains in work-hardening plastic solids. *J. Appl. Mech*. **23**; 493–496 (1956).

[9] Ziegler H. A modification of Prager's hardening rule. *Q. Appl. Math*. **17**; 55–60 (1959).

[10] Hencky H. Zür theorie plastischer deformationen und der hiedurch im material hervorgerufenen nebenspannungen. *Zeits. Ang. Math. Mech*. **4**; 323–334 (1924).

[11] Femi E. *Thermodynamics*. Dover, New York (1936).

[12] Drucker DC. A definition of stable inelastic material. *J. Appl. Mech*. **26**; 101–106 (1959).

[13] Washizu K. *Variational methods in elasticity and plasticity*. Pergamon Press, Oxford (1975).

[14] Nayak GC, Zienkiewicz OC. Convenient forms of stress invariants for plasticity. *J. Struct. Div. ASCE*, **98**; 949–953 (1972).

[15] Lode W. Versuche über den einfluss der mittleren hauptspannung auf das fliessen der metalle eisen, kupfer und nickel. *Zeits. Physik*, **36**; 913–939 (1926).

[16] Ellis R, Gulick D. *Calculus with analytic geometry*. Harcourt Brace Jovanovich, New York (1978).

[17] Gudehus G. Elastoplastische stoffgleichungen für trockenen sand. *Ing. Arch*. **42**; 151–169 (1973).

[18] Argyris JH, Faust G, Szimmat J, Warnke EP, Willam KJ. Recent developments in the finite element analysis of prestressed concrete reactor vessels. *Nucl. Eng. Des*. **28**; 42–75 (1974).

[19] Willam KJ, Warnke EP. Constitutive model for triaxial behavior of concrete. *Colloquium on Concrete Structures Subjected to Triaxial Stresses,* ISMES Bergamo, IABSE Report, **19** (1974).

[20] Jiang J, Pietruszczak S. Convexity of yield loci for pressure sensitive materials. *Comput. Geotech*. **5**; 51–63 (1988).

[21] Pietruszczak S, Jiang J, Mirza FA. An elastoplastic constitutive model for concrete. *Int. J. Solids Struct.* **24**; 705–722 (1988).

[22] Chen WF. *Plasticity in reinforced concrete.* McGraw-Hill Inc., New York (1982).

[23] MacGinley TJ, Choo BS. *Reinforced concrete: design theory and examples.* Taylor & Francis Group, New York (1990).

[24] Schofield A, Wroth P. *Critical state soil mechanics.* McGraw Hill, Maidenhead (1968).

[25] Muir Wood D. *Soil behaviour and critical state soil mechanics.* Cambridge University Press, Cambridge (1990).

[26] Pietruszczak S, Poorooshasb HB. Modelling of cyclic behaviour of soils. In: *Developments in Soil Mechanics and Foundation Engineering* – 2 (Eds. Banerjee, Butterfield), Chapter 6. Elsevier Applied Science, 139–184 (1985).

[27] Wilde P. Two-invariants dependent model of granular media. *Arch. Mech.* **29**; 799–809 (1977).

[28] Nova R. On the hardening of soils. *Archiwum Mechaniki Stosowanej* **29**; 445–458 (1977).

[29] Mroz Z, Norris VA, Zienkiewicz OC. An anisotropic hardening model for soils and its application to cyclic loading. *Int. J. Numer. Anal. Methods Geomech.* **2**; 203–221 (1978).

[30] Mroz Z. On the description of anisotropic workhardening. *J. Mech. Phys. Solids*, **15**; 163–175 (1967).

[31] Dafalias YF, Popov EP. A model for non-linearly hardening materials for complex loadings. *Acta Mechanica* **21**; 173–192 (1975).

[32] Dafalias YF, Popov EP. Cyclic loading for materials with a vanishing elastic region. *Nucl. Eng. Des.* **41**; 293–302 (1977).

[33] Krieg RD. A practical two-surface plasticity theory. *J. Appl. Mech.* **42**; 641–646 (1975).

[34] Mroz Z, Norris VA, Zienkiewicz OC. Application of an anisotropic hardening model in the analysis of elasto-plastic deformation of soils. *Geotechnique* **29**; 1–34 (1979).

[35] Pietruszczak S, Mroz Z. On hardening anisotropy of K_0-consolidated clays. *Int. J. Numer. Anal. Methods Geomech.* **7**; 19–38 (1983).

[36] Mroz Z, Pietruszczak S. A constitutive model for sand with anisotropic hardening rule. *Int. J. Numer. Anal. Methods Geomech.* **7**; 305–320 (1983).

[37] Poorooshasb HB, Pietruszczak S. On yielding and flow of sand: a generalized two-surface model. *Comput. Geotech.* **1**; 33–58 (1985).

[38] Pietruszczak S, Stolle DFE. Modelling of sand behaviour under earthquake excitation. *Int. J. Numer. Anal. Methods Geomech.* **11**; 221–240 (1987).

[39] Simo JC, Hughes TJR. *Computational inelasticity.* Springer Verlag, New York (1998).

[40] Simo JC, Hughes TJR. General return mapping algorithms for rate-independent plasticity. In: *Constitutive Laws for Engineering Materials: Theory and Applications* **1**. Elsevier; 221–232 (1987).

[41] Simo JC, Ortiz M. A unified approach to finite deformation elastoplastic analysis based on the use of hyperelastic constitutive equations. *Comput. Methods. Appl. Mech. Eng.* **49**; 221–245 (1985).

[42] Ortiz M, Simo JC. An analysis of a new class of integration algorithms for elastoplastic constitutive relations. *Int. J. Numer. Methods. Eng.* **23**; 353–366 (1986).

[43] Gvozdev AA. The determination of the value of the collapse load for statically indeterminate systems undergoing plastic deformation. *Int. J. Mech. Sci.* **1**; 322–335 (1960). (Translation from Russian of the 1936 paper in the Proceedings of the Conf. on Plastic Deformations, Academy of Sciences USSR).

[44] Drucker DC, Prager W, Greenberg HJ. Extended limit design theorems for continuous media. *Q. Appl. Math.* **9**; 381–389 (1952).

[45] Hodge PG. *The plastic analysis of structures*. McGraw-Hill, New York (1959).

[46] Nielsen MP. *Limit analysis and concrete plasticity*. Taylor and Francis, CRC Press, New York (1998).

[47] Chen WF. *Limit analysis and soil plasticity*. Elsevier Science, Amsterdam (1975).

[48] Chen WF, Liu X. *Limit analysis in soil mechanics*. Elsevier Science, Amsterdam (1990).

[49] Salencon J. *Calcul à la rupture et analyse limite*. Presses des Ponts et Chaussées, Paris (1983).

[50] Prandtl L. Über die härte plastischer körper. Nachrichten von der Gesellschaft der Wissenschaften zu Göttingen. *Mathematisch-Physikalische Klasse* **12**; 74–85 (1920).

[51] Drucker DC, Prager W. Soil mechanics and plastic analysis or limit design. *Q. Appl. Math.* **10**; 152–165 (1952).

[52] Pietruszczak S, Mroz Z. On failure criteria for anisotropic cohesive-frictional materials. *Int. J. Numer. Anal. Methods Geomech.* **25**; 509–524 (2001).

[53] Pietruszczak S, Mroz Z. Formulation of anisotropic failure criteria incorporating a microstructure tensor. *Comput. Geotech.* **26**; 105–112 (2000).

[54] Duveau G, Shao JF, Henry JP. Assessment of some failure criteria for strongly anisotropic materials. *Mech. Cohesive Frictional Mater.* **3**; 1–26(1998).

[55] Hill R. A theory of the yielding and plastic flow of anisotropic metals. *Proc. R. Soc. Lond.* **193**; 281–297 (1948).

[56] Tsai SW, Wu E. A general theory of strength of anisotropic materials. *J. Compos. Mater.* **5**; 58–80 (1971).

[57] Boehler JP, Sawczuk A. Equilibre limite des sols anisotropes. *J. de Mecanique* **3**; 5–33 (1970).

[58] Boehler JP, Sawczuk A. On yielding of oriented solids. *Acta Mechanica* **27**; 185–206 (1977).

[59] Cowin SC. Fabric dependence of an anisotropic strength criterion. *Mech. Mater.* **5**; 251–260 (1986).

[60] Kanatani K. Distribution of directional data and fabric tensor. *Int. J. Eng. Sci.* **22**; 149–161 (1984).

[61] Nova R, Zaninetti A. An investigation into the tensile behavior of a schistose rock. *Int. J. Rock Mech. Min. Sci.* **27**; 231–242 (1990).

[62] Azami A, Pietruszczak S, Guo P. Bearing capacity of shallow foundations in transversely isotropic granular media. *Int. J. Numer. Anal. Methods Geomech.* **34**; 771–793 (2010).

[63] Pande GN, Sharma KG. Multilaminate model of clays – a numerical evaluation of the influence of rotation of the principal stress axes. *Int. J. Numer. Anal. Methods Geomech.* **7**; 397–418 (1983).

[64] Pietruszczak S, Pande GN. Description of soil anisotropy based on multi-laminate framework. *Int. J. Numer. Anal. Methods Geomech.* **25**; 197–206 (2001).

[65] Kolymbas D, Wu W. Recent results of triaxial tests with granular materials. *Powder Technol.* **60**; 99–119 (1990).

[66] Lade PV. Elastic-plastic stress-strain theory for cohesionless soil with curved yield surfaces. *Int. J. Solids Struct.* **13**; 1019–1035 (1977).

[67] Elliott GM, Brown ET. Yield of a soft, high porosity rock. *Geotechnique* **35**; 413–423 (1985).

[68] Xie SY, Shao JF. Elastoplastic deformation of a porous rock and water interaction. *Int. J. Plast.* **22**; 2195–2225 (2006).

[69] Lade PV. Modelling the strengths of engineering materials in three dimensions. *Mech. Cohesive-Frictional Mater.* **2**; 339–356 (1997).

[70] Akai K, Mori H. Ein versuch der bruchmecanismus von sandstein under mehrachsigen spannungszustand. *Proc. 2nd Intern. Congr. Rock Mech.* Belgrade, **2**; 3–30 (1970).

[71] Lade PV, Duncan JM. Cubical triaxial tests on cohesionless soil. *J. Soil Mech. Found. Div. ASCE* **99**; 793–812 (1973).

[72] Yamada Y, Ishihara K. Anisotropic deformation characteristics of sand under three dimensional stress conditions. *Soils Found.* **19**; 79–94 (1979).

[73] Haimson B, Chang C. Brittle fracture in two crystalline rocks under true triaxial compressive stresses. In: *Petrophysical Properties of Crystalline Rocks* (Eds. Harvey, Brewer, Pezard, Petrov). *Geological Society London Special Publication* **240**; 47–59 (2005).

[74] Khalid A, Alshibli A, Sture S. Shear band formation in plane strain experiments of sand. *J. Geotech. Geoenviron. Eng.* **126**; 495–503 (2000).

[75] Pietruszczak S, Mroz Z. Finite element analysis of strain softening materials. *Int. J. Numer. Methods. Eng.* **17**; 327–334 (1981).

[76] Bazant ZP, Pijaudier-Cabot G. Nonlocal continuum damage, localization instability and convergence. *J. Appl. Mech.* **55**; 287–293 (1988).

[77] Bazant ZP, Lin FB. Non-local yield limit degradation. *Int. J. Numer. Methods Eng.* **26**; 1805–1823 (1988).

[78] Muhlhaus HB, Vardoulakis I. The thickness of shear bands in granular materials. *Geotechnique* **37**; 271–283 (1987).

[79] Triantafyllidis N, Aifantis EC. A gradient approach to localization of deformation. Part I: Hyperelastic materials. *J. Elast.* **16**; 225–237 (1986).

[80] de Borst R, Muhlhaus HB. Gradient-dependent plasticity: formulation and algorithmic aspects. *Int. J. Numer. Methods. Eng.* **35**; 521–539 (1992).

[81] Pietruszczak S, Xu G. Brittle response of concrete as a localization problem. *Int. J. Solids Struct.* **32**; 1517–1533 (1995).

[82] Pietruszczak S. On homogeneous and localized deformation in water-infiltrated soils. *J. Damage Mech.* **8**; 233–253 (1999).

[83] Baladi GY, Rohani B. An elastic-plastic constitutive model for saturated sand subjected to monotonic and/or cyclic loadings. *Proc. 3rd Intern. Conf. Num. Meth. Geomech.* Aachen, **1**; 389–404 (1979).

[84] Castro G. Liquefaction of sands. *Harvard University Soil Mechanics Series.* Pierce Hall, Cambridge, MA No. 81 (1969).

[85] Coussy O. *Mechanics of Porous Continua.* Wiley and Sons, Chichester (1995).

[86] Sulem J, Ouffroukh H. Hydromechanical behaviour of Fontainbleau sandstone. *Rock Mech. Rock Eng.* **39**; 185–213 (2006).

[87] Tatsuoka F, Ishihara K. Drained deformation of sand under cyclic stress reversing direction. *Soils Found.* **14**; 51–94 (1974).

[88] Ishihara K, Tatsuoka F, Yasuda S. Undrained deformation and liquefaction of sand under cyclic stresses. *Soils Found.* **15**; 29–44 (1975).

[89] Niandou H, Shao JF, Henry JP, Fourmaintraux D. Laboratory investigation of the mechanical behaviour of Tournemire shale. *Int. J. Rock Mech. Min. Sci.* **34**; 3–16 (1977).

[90] Calvetti F, Tamagnini C, Viggiani G. On the incremental behaviour of granular soils. In: *Numerical Models in Geomechanics NUMOG VIII (Eds.* Pande, Pietruszczak). A.A. Balkema Publication; 3–9 (2002).

[91] Koiter WT. Stress-strain relation, uniqueness and variational theorems for elastic-plastic materials with a singular yield surface. *Q. Appl. Math.* **11**; 350–354 (1953).

Appendix

Suggested exercises

(i) Numerical assignments

1. Write a numerical code for integration, in the 'triaxial' $\{p, q\}$ space, of the deviatoric hardening model (Section 3.3.1). Use explicit Euler's scheme and examine both stress and strain-controlled paths (viz. Table 5.1 and 5.2 in Section 5.2).
   - Verify the code using the results of simulations provided in Section 3.3.2 (Figures 3.15 and 3.16). Employ the same material parameters and testing conditions, viz. drained and undrained deformation.
   - Perform a parametric study investigating the influence of (i) confining pressure, (ii) the value of constant A in eq.(3.62) and (ii) ratio of η_c/η_f on the material response.
   - Predict the response under cyclic undrained axial compression. Use the testing conditions and material properties the same as those employed in the simulations given in Figures 4.6 and 4.7 in Section 4.2.2. Comment on the quality of these predictions.

   (Note: the numerical algorithm given in Table 5.2 of Section 5.2 needs to be enhanced to incorporate loading/unloading criteria that will ensure that for the reverse loading the response remains elastic. Also, the constitutive framework described in Section 3.3.1 needs to be extended to include the extension ($q < 0$) regime.)

2. Complete the assignment #1 using the Critical State model (Section 3.2.1). Assume the material parameters and testing conditions analogous to those employed in the examples given in Section 3.3.2.
   - Verify your predictions against the results of simulations in Figures 3.11–3.12.
   - Perform a parametric study investigating the influence of compression and swelling indexes λ, κ on the undrained response.
   - For cyclic tests use the testing conditions corresponding to simulations given in Figures 4.2 and 4.4 in Section 4.1.2.

3. Write a general 3D strain-controlled code, based on explicit Euler's scheme, for integration of the constitutive relation incorporating deviatoric hardening. Use the framework of Section 3.3.3. Verify the code based on the results of simulation of 'triaxial' tests of assignment #1. Predict the material response under the following loading histories:
   (i) $d\varepsilon_{12} > 0$; remaining components of $d\varepsilon_{ij}$ equal to zero
   (ii) $d\varepsilon_{12} > 0$, $d\varepsilon_{13} = 0$, $d\varepsilon_{23} = 0$; $d\varepsilon_{11} = -d\varepsilon_{12}$, $d\varepsilon_{22} = d\varepsilon_{33} = d\varepsilon_{12}/2$

4. Complete the assignment #3 using the Critical State model (Section 3.2.3).

(ii) Other brief exercises

1. Evaluate the derivatives: $\dfrac{\partial \sigma_m}{\partial \sigma_{ij}}, \dfrac{\partial \overline{\sigma}}{\partial \sigma_{ij}}, \dfrac{\partial J_3}{\partial \sigma_{ij}}$

Answer: See eq.(2.66) of Chapter 2

2. Consider an elastic – perfectly plastic Drucker-Prager material for which $F = \overline{\sigma} - \eta\,\sigma_m = 0$ and $\psi = \overline{\sigma} - \mu\,\sigma_m = const.$ For an active loading process, define the admissible range of values of μ so that the rate of energy dissipation is positive.

Answer: $\mu \leq \eta$

3. Consider an elastoplastic strain-hardening material for which $f(\sigma_{ij}, \varepsilon^p_{ij}) = 0$ and $\psi = g(\sigma_{ij}) - const. = 0$.
   (i) Specify the form of the compliance tensor C_{ijkl} in the incremental relation $d\varepsilon_{ij} = C_{ijkl}\,d\sigma_{kl}$
   (ii) Define the active loading, neutral loading and unloading process for this material

Answer: (i) $C_{ijkl} = C^e_{ijkl} + H_p^{-1}\left(\dfrac{\partial g}{\partial \sigma_{ij}}\dfrac{\partial f}{\partial \sigma_{kl}}\right)$; $H_p = -\dfrac{\partial f}{\partial \varepsilon^p_{ii}}\dfrac{\partial g}{\partial \sigma_{pp}}$; (ii) See eqs.(1.9)–(1.11) of Chapter 1

4. (i) Derive the expression for the rate of energy dissipation in a von Mises material $F = \overline{\sigma} - k = 0$, where $k = $ const. Express the result in terms of the plastic multiplier.

   (ii) Assuming

$$f = \overline{\sigma} - k = 0; \quad k = k_0 + (k_f - k_0)\frac{\varepsilon^p}{A + \varepsilon^p}; \quad d\varepsilon = \left(\frac{1}{2} de_{ij} de_{ij}\right)$$

   Derive the expression for plastic hardening modulus H_p. Define the condition for which $H_p \to 0$. What is the corresponding value of k?

Answer: (i) $dW_p = d\lambda\, k$;

$$\text{(ii) } H_p = \frac{1}{2}\frac{dk}{d\varepsilon^p} = \frac{(k_f - k_0)A}{2(A + \varepsilon^p)^2}; \quad H_p \to 0 \quad for \quad \varepsilon^p \to \infty \quad \Rightarrow \quad k \to k_f$$

5. Derive an expression for the constitutive matrix for an elastic – perfectly plastic material. Assume $F(\boldsymbol{\sigma}) = 0$, $\psi(\boldsymbol{\sigma}) = 0$ as the general equations of the failure and plastic potential surfaces.

Answer: See eq.(2.72) of Chapter 2

6. Using p,q invariants, define the general form of the constitutive matrix $[D]$ in the relation $d\boldsymbol{\sigma} = [D]\,d\boldsymbol{\varepsilon}$. Assume $F = f(p,q) - g(\varepsilon^p_v) = 0$ and $\psi = f = const.$

Answer: See eq.(3.28) of Chapter 3

7. Two saturated samples of *overconsolidated* kaolin were subjected to (i) drained and (ii) undrained axial compression at the initial confining pressures of 40 kPa and 60 kPa, respectively. In the drained test, the sample failed by applying *additional* vertical stress of 100 kPa, whereas in the undrained test the increase in vertical stress was 60 kPa and the measured excess of pore pressure at failure was 40 kPa. Draw the corresponding stress paths in $\{p, q\}$ space and determine the material parameters for the Mohr-Coulomb criterion.

Answer: $\eta_f = 1.2$, $\mu_f = 12\,\text{kPa} \Rightarrow \phi = 30^0$, $c = 5.77\,\text{kPa}$ (cf. eq.(3.8) of Chapter 3)

8. Two 'triaxial' tests were performed on identical samples of *loose sand* at the initial confining pressure of 60 kPa:
   (i) undrained axial compression; the sample failed at $q = 60\,\text{kPa}$ developing the excess of pore pressure of 30 kPa, and (ii) undrained lateral extension.
   Plot the corresponding stress paths in $\{p, q\}$ space and determine the excess pore pressure at failure for test (ii). What are the values of the angle of internal friction and cohesion?

Answer: $p_w = -30\,\text{kPa}$; $\phi = 30°$, $c = 0$ (granular material)

9. Assuming $F = f(p, q) - g(\varepsilon_q^p) = 0$ and $\psi = f = const.$:
   (i) derive the expression for the plastic hardening modulus H_p.
   (ii) Define $F = 0$ and $\psi = const.$ for von Mises material and specify the conditions required for $H_p = 0$.
   (iii) Define the general form of the compliance matrix in the relation between strain and stress increments.

Answer: (i) $H_p = \dfrac{dg}{d\varepsilon_q^p}\dfrac{\partial f}{\partial q}$, (ii) $F = q - g(\varepsilon_q^p) = 0$, $\psi = q - const = 0$, $\dfrac{dg}{d\varepsilon_q^p} \to 0$

(iii) See eq.(3.60) of Chapter 3 under $\psi = f = const.$

10. For a rigid-plastic material with
    (i) deviatoric hardening : $f = q - \eta(\varepsilon_q^p)p = 0$, $\psi = q + \eta_c p \ln(p/\bar{p}) = 0$
    (ii) volumetric hardening: $f(p, q, e^p) = (p - a)^2 + (q/\eta_f)^2 - a^2 = 0$; $a = a(e^p)$
    determine the stress trajectory associated with in-situ (K_0) stress conditions. Is the solution admissible?
    Note: K_0-conditions are associated with no lateral deformation, i.e. $\varepsilon_2 = \varepsilon_3 = 0$

Answer: (i) $\eta = \eta_c - 3/2 = const$. Not admissible as $\eta = const$ defines a neutral path.

(ii) $\eta = q/p = \frac{1}{2}(\sqrt{9 + 4\eta_f^2} - 3)$; admissible

11. Consider an isotropic strain-hardening material with intersecting loading surfaces:

$$f_1 = q - \eta(\varepsilon_q^p)p = 0, \quad f_2 = p - a(\varepsilon_v^p) = 0$$

Assuming Koiter's-type [91] of associated flow rule:

$$d\boldsymbol{\varepsilon}^p = d\lambda_1 \frac{\partial f_1}{\partial \boldsymbol{\sigma}} + d\lambda_2 \frac{\partial f_2}{\partial \boldsymbol{\sigma}}$$

define the ratio $\zeta = d\varepsilon_q^p / d\varepsilon_v^p$. Specify the value of ζ for a standard single surface formulation based on $f_1 = 0$ and $f_2 = 0$, respectively.

Answer: $\zeta = \frac{a'}{p\eta'}\left(-\eta + \frac{dq}{dp}\right)$, where a', η' are the derivatives with respect to plastic strain components. Note that the direction of plastic flow, viz. ζ, depends on the direction of stress increment. For $f_1 = 0$ there is $\zeta = -1/\eta'$, while $\zeta = 0$ for $f_2 = 0$.

Milton Keynes UK
Ingram Content Group UK Ltd.
UKHW051854071024
449327UK00025B/1950

9 780367 577148